The authors present a basic and accessible introduction to the world of microbiology. In three chapters, this book provides both a foundation and overview of the subject. In the first chapter, 'Microbial structure and mode of life', the structure and functioning of fungi, bacteria and viruses are discussed (with particular attention being paid to their description and discussion of their reproduction and nutrition). The second chapter, 'Handling microbes' introduces the methods used to culture, control and study these organisms in the laboratory. The final chapter covers the 'Isolation, classification and identification of microbes'.

This book is essential reading for anyone becoming interested in this subject, whether it be sixth form students, their teachers, or undergraduates.

D0139589

Introductory microbiology

The Institute of Biology aims to advance both the science and practice of biology. Besides providing the general editors for this series, the Institute publishes two journals *Biologist* and the *Journal of Biological Education*, conducts examinations, arranges national and local meetings and represents the views of its members to government and other bodies. The emphasis of the *Studies in Biology* will be on subjects covering major parts of first-year undergraduate courses. We will be publishing new editions of the 'bestsellers' as well as publishing additional new titles.

Titles available in this series

An Introduction to Genetic Engineering, D. S. T. Nicholl

Photosynthesis, 5th edition, D. O. Hall and K. K. Rao

Introductory microbiology

J. Heritage, E. G. V. Evans and R. A. Killington

Department of Microbiology, University of Leeds

Published in association with the Institute of Biology

CAMBRIDGE
UNIVERSITY PRESS

Published by the Press Syndicate of the University of Cambridge
The Pitt Building, Trumpington Street, Cambridge CB2 1RP
40 West 20th Street, New York, NY 10011-4211, USA
10 Stamford Road, Oakleigh, Melbourne 3166, Australia

First published 1996

Printed in Great Britain at the University Press, Cambridge

A catalogue record for this book is available from the British Library

Library of Congress cataloguing in publication data

Heritage, J.
    Introductory microbiology / J. Heritage, E. G. V. Evans and R. A.
Killington.
      p.   cm – (Studies in biology)
    Includes index.
    ISBN 0 521 44516 7
    1. Microbiology.   I. Evans, E. Glyn V.   II. Killington, R. A.
III. Series.
QR41.2.H46   1996
576 – dc20   95-10245 CIP

ISBN 0 521 44516 7 hardback
ISBN 0 521 44977 4 paperback

SE

Contents

General preface to the series

Charged by its Royal Charter to promote biology and its understanding, the Institute of Biology recognises that it is not possible for any one text book to cover the entirety of a course. If evidence was needed, the success of the *Studies in Biology* series was a testimony to the need for specialist, up-to-date publications in education. The Institute is therefore pleased to collaborate with Cambridge University Press in producing a new title in the *Studies in Biology* series.

The new series is set to provide as great a boon to the new generation of students as the original did to their parents.

Suggestions and comments from readers will always be welcomed and should be addressed either to the Studies in Biology Editorial Board at Cambridge University Press or c/o The Books Committee at the Institute.

Robert Priestley
The General Secretary

The Institute of Biology
20–22 Queensberry Place
London SW7 2DZ

Preface

When I was first approached to write this book for the *Studies in Biology* series, I was delighted. As a student this series had a profound influence on me, and I was an avid reader of the books. They provided an easily affordable access to a very wide range of topics in biology, and I still use one or two books from my student days.

I was also very daunted at the prospect of writing such a short book to cover so vast a topic as microbiology. It is a subject that impinges on almost every aspect of human existence. To attempt to cover the whole of the subject would have been an impossible task. Consequently, some very difficult decisions have had to be made, and some very interesting material has had to be omitted. It is to be regretted that we could not expand upon topics such as the story of the near collapse of Winchester Cathedral as a result of fungal decomposition of the oak raft that supported the structure, following drainage of nearby farm-land.

My co-authors and I have confined ourselves to a consideration of the aspects of microbiology in which we have research experience: bacteriology, mycology and virology. We have also tried to concentrate on the fundamental problems in the subject. What constitutes microbes? How do they differ from higher organisms and from each other? How can microbes be controlled, visualised, enumerated and cultured? How are they classified? The answers to these and other questions can be found within our text.

Our approach is derived from Introductory Microbiology courses that we have successfully taught in Leeds for a number of years. These have provided many students with a firm foundation on which to build further studies in the subject, and have also provided a solid framework of microbiological concepts for students whose requirements are more modest.

During the preparation of this book, I have sought the help and advice of a number of people to whom I am indebted. My Head of Department, Keith Holland, has been supportive throughout the project. He has read much of the text critically, and made many excellent suggestions for improving the book. Deborah Gascoyne-Binzi and Anna Snelling have likewise been a source of critical support regarding the structure and content of the book. Brian Hartley has advised me on the Latin and Greek derivations of some of the microbiological terminology. At various stages in the project, my colleagues have commented on aspects of the manuscript. These have included David Adams, Ian Halliburton, Peter Hawkey, Pat Hayes and John Shoesmith. Since this book is intended for a non-specialist audience, I am also grateful for the advice from Dunstant Adams and Leon Carberry. My sincere thanks are also extended to Bill White. To all these, and to others I owe a debt of gratitude. Their advice has been invaluable, but any mistakes in the text are now all mine. I am also grateful for the support of my publishers, and in particular I would like to thank Alan Crowden and Tim Benton. I am very grateful for their tolerance and patience.

Finally I would like to express my sincere thanks to my wife Pauline and to my children, Janet and Tom. This book was written during a period of rapid expansion in Higher Education, and when the University of Leeds was embracing modularisation of its courses. Writing this book provided an enjoyable relief, but there have been times recently when my family must have thought that I had regressed to become a somewhat irritable bald-patch that sat in a quiet corner of the living room surrounded by a mountain of papers!

JH
York

1

Microbial structure and mode of life

1.1   Introduction

Early biologists found it convenient to classify all living things as either animals or plants. To many people today, this grouping still seems perfectly adequate. However, examination of the life-forms that exist on Earth shows that this classification is unsatisfactory. Although there is a superficial resemblance between green plants and the fungi, these two groups are divided by profound biological differences. Unlike green plants, fungi cannot manufacture their own food from water and carbon dioxide by the process of photosynthesis. Rather, they require a supply of organic matter from which they can derive their energy. Fungal cellular composition is dissimilar from that of green plants, and the structural polymers of their **cell walls** are markedly different. **Fungi** are therefore now accorded their own status as a third kingdom. Furthermore, for many years the classification of microscopic organisms proved to be difficult. Photosynthetic microbes behave very differently from higher plants. It was therefore proposed towards the end of the nineteenth century that microscopic life forms should be classified as a fourth kingdom. This was the kingdom **Protista**, proposed in 1866, at a time when the scientific study of microbiology was in its infancy. This was, however, almost 200 years after Antonie van Leeuwenhoek described '**animalcules**' following his development of the optical microscope.

During the twentieth century there have been many advances in microscopy, including the development of the electron microscope. This has enabled subcellular structures to be studied in great detail and has revealed that the Protista may be divided into two major groups. Primitive microorganisms such as **bacteria** lack a clearly defined, membrane-bound **nucleus** and are called **prokaryotes**. This word is derived from two Greek words; *pro*, meaning before, and *karyon*, a kernel. Prokaryotes are organisms that evolved before the cell's nucleus, its kernel, was properly developed. More advanced microscopic life-forms, as well as having a variety of subcellular **organelles**, also possess a proper membrane-bound nucleus. These are referred to as **eukaryotes** because they have a true nucleus (Greek: *eu-*, well, true or easy). Biologists now reserve the kingdom Protista for eukaryotic microbial life-forms and separate prokaryotes into their own kingdom, the **Monera**. *Monera* is 'new' Latin for non-nucleated protoplasmic masses. Some microbiologists prefer to refer to this kingdom as the **Prokaryotes**. Thus, life-forms may be classified into five kingdoms: Animals, Plants, Fungi, Protista, and Monera or Prokaryotes. Table 1.1 indicates the characteristic features of members of the different kingdoms.

Microbiology is the study of microscopic life-forms including the microscopic fungi and Protista as well as prokaryotes. It also encompasses a study of **viruses**: subcellular structures comprising nucleic acid inside a protein coat and sometimes covered in a membrane. Viruses cannot replicate outside a host cell, and thus represent the ultimate parasites. Despite their limitations, viruses have managed to parasitise animal, plant and even bacterial cells. In a book of this size, it is not possible to cover the whole of microbiology, and so the text is devoted to a description of the biology of bacteria (prokaryotes), viruses and the microfungi.

1.2   A comparison of eukaryotic and prokaryotic cells

The cells of eukaryotic organisms have a much more complex structure than do prokaryotic cells. Eukaryotic cells have a highly developed internal membrane and microtubular structure, together with subcellular organelles. Such structures are absent from many prokaryotes, although some prokaryotes do have specialised internal membranes associated with particular metabolic functions. The nucleus of the eukaryotic cell is membrane bound. Associated with the nucleus is the **centriole**, a structure used in the regulation of **mitosis**. This is the process of cell division that ensures that each daughter cell receives a complete set of **chromosomes**. Protein synthesis in eukaryotic cells is carried out on **ribosomes** associated with the **endoplasmic reticulum**, a network of internal membranes. The endoplasmic reticulum is contiguous with the **Golgi apparatus**; a structure associated with the export of proteins (Fig. 1.1).

Table 1.1 *Characteristic features of members of the five kindgoms*

	Animals	Plants	Fungi	Protista	Prokaryotes
Multicellular	Yes	Yes	Yes	No	No
Membrane-bound nucleus	Yes	Yes	Yes	Yes	No
Nuclear material present as naked DNA	No	No	No	No	Yes
Cytoplasmic membrane-bound organelles	Yes	Yes	Yes	Yes	No
Cell walls	No	Yes (mainly cellulose-based)	Yes (mainly chitin-based)	Some	The vast majority. Mostly peptido-glycan-based, but other polymers are found
Growth can be achieved using simple chemicals	No	Yes	No	Some	Some
Growth requires a supply of complex organic molecules	Yes	No	Yes	Some	Some
Chloroplasts	No	Yes	No	Some	No
Mitochondria	Yes	Yes	Yes	Yes	No
Ribosomes	Larger cytoplasmic, smaller mitochondrial	Larger cytoplasmic, smaller mitochondrial and chloroplast	Larger cytoplasmic, smaller mitochondrial	Larger cytoplasmic, smaller mitochondrial and chloroplast (if present)	Small

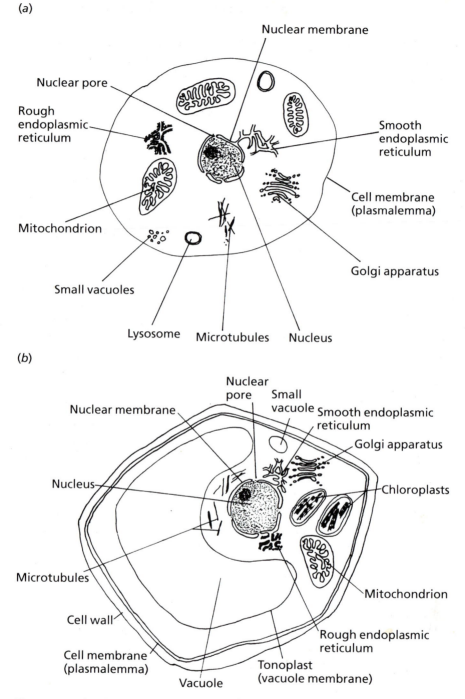

Fig. 1.1. Stylised eukaryotic cells. (a) A stylised animal cell and (b) a stylised plant cell.

Cellular respiration, generating ATP, is carried out in **mitochondria**. These are organelles delineated by a double membrane. The mitochondrial inner membrane carries the respiratory enzymes or **cytochromes**, and is highly convoluted to form **cristae**. These greatly increase the surface area of the mitochondrial inner membrane, the surface at which ATP generation occurs. Plant cells, in addition, carry **plastids**. These are also organelles bounded by a double membrane, and in green plant tissues plastids become specialised as **chloroplasts**, which are also involved in the generation of ATP. Chloroplasts generate ATP by the process of **photosynthesis** whereas mitochondria generate ATP by aerobic respiration. Chloroplasts have elaborate internal membrane structures called **thylakoids**. Both mitochondria and chloroplasts have their own DNA and are capable of synthesising particular proteins. This process is much more akin to bacterial protein synthesis than to cytoplasmic protein synthesis in eukaryotic cells. Many biologists believe that mitochondria and chloroplasts represent the descendants of obligate intracellular endosymbiotic bacteria that date from the earliest days of eukaryotic cells. The ancestral **endosymbionts** are thought to have evolved to lose most of their bacterial structure. However, mitochondria and chloroplasts retain a genetic apparatus that provides them with the ability to synthesise certain of their own proteins. They also provide metabolic processes that have proved to be of great value to the eukaryotic host (Fig. 1.2).

Both plant and animal cells have membrane-bound **vacuoles**. Plant cell vacuoles are large and bounded by a special membrane called a **tonoplast**. Animal cells contain smaller, more numerous vacuoles. Additionally, animal cells contain **lysosomes**; subcellular sacs filled with lytic enzymes. These may fuse with vacuoles containing foreign bodies to form **phagosomes**. Inside the phagosome, lytic enzymes digest foreign material, and the breakdown products may be absorbed into the cell. Cytoplasmic streaming in eukaryotic cells is the process by which metabolites are distributed through the cell. It is controlled by a system of microtubules. Some eukaryotic cells elaborate (or develop) organelles of motility that extend into the environment. Short structures are referred to as **cilia**; longer organelles are called **flagella**. Eukaryotic cilia and flagella have a characteristic internal structure comprising nine pairs of microtubules encircling a further two pairs of tubules. This is often referred to as a 9 + 2 structure, and is confined to eukaryotic cells.

Eukaryotic cells are bounded by a cytoplasmic membrane, the **plasmalemma**. In animal cells, the plasmalemma is the outermost structure; plant cells have a **cellulose**-based cell wall outside the plasmalemma to provide additional protection. Eukaryotic membranes are stabilised by the presence of sterols.

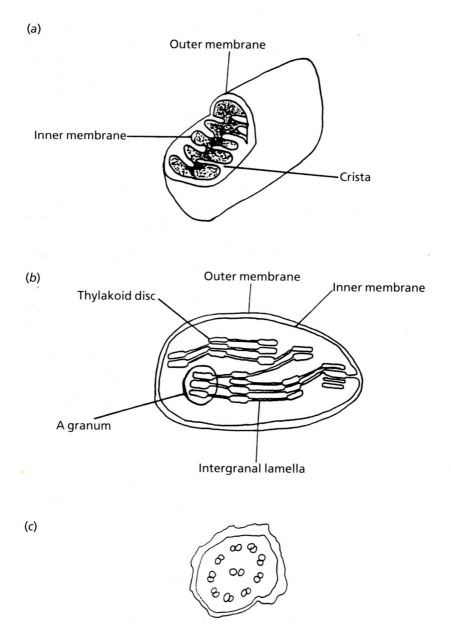

Fig. 1.2.   Some subcellular structures found in eukaryotic cells. (*a*) A cut-away diagram of a mitochondrion, (*b*) section through a chloroplast, and (*c*) cross-section through a eukaryotic flagellum, showing the typical '9 + 2' structure.

In contrast to eukaryotic cells, prokaryotes have a very simple internal structure. As their name implies, they do not have a clearly defined nucleus delineated by a nuclear membrane. Rather, the chromosome forms a densely packed area within the cytoplasm known as the **nucleoid**. Extrachromosomal DNA structures, at one time called **episomes** and now more commonly known as **plasmids**, also exist within the cytoplasm. The prokaryotic ribosomes, which are smaller than those found in eukaryotic cells, are not attached to internal membranes. Rather, multiple ribosomes line up along messenger RNA molecules to form structures called **polysomes**.

The size of ribosomes is conventionally expressed as the rate at which they sediment in a density gradient such as a sucrose gradient. The unit of measurement is the Svedberg unit. Eukaryotic cytoplasmic ribosomes are described as 80 S ribosomes, and these are larger than prokaryotic 70 S ribosomes. Ribosomes are made up of two subunits. In eukaryotic cells these are 60 S and 40 S subunits, whereas in prokaryotes there are 50 S and 30 S subunits. The sedimentation rate of a body depends upon its shape as well as its density. Thus the whole ribosome unit has a lower sedimentation rate than that of its component subunits. The ribosomes found in chloroplasts and mitochondria are much closer in size, and behaviour, to the 70 S prokaryotic ribosome than to the 80 S eukaryotic ribosome. This observation lends weight to the endosymbiotic theory for the origin of mitochondria and chloroplasts.

Archaebacteria represent some of the most ancient of life-forms that persist today. They can be found in some of the most extreme habitats on Earth. For example archaebacteria have been isolated from coal slag heaps, in an environment with a temperature of about 60 °C, and **pH** of about 1, and they can be found around underwater volcanoes. They may be differentiated from other bacteria, the eubacteria, using a variety of characteristics, especially with respect to ribosomal and membrane structures, but these differences are not sufficient to warrant the separation of archaebacteria and eubacteria into two separate kingdoms. The prefix archae- comes from the Greek word *arkhaio*, meaning ancient. **Eubacteria** are the most familiar and commonly studied of bacteria.

The cytoplasm of prokaryotic cells is bounded by a cell membrane, and many bacteria have a cell wall outside the cell membrane. The principal cell wall polymer of archaebacteria varies from group to group, whereas in eubacteria the main component is **peptidoglycan**. Various structures that act in cellular adhesion and locomotion are associated with the prokaryotic cell wall. Many prokaryotes have an additional outer membrane beyond the cell wall, and some bacteria have external structures beyond this.

1.3 Morphology of fungi

The study of fungi is referred to as mycology, and is one of the oldest disciplines in microbiology. Fungi are amongst the most generally familiar organisms that are studied in microbiology. Everyone will recognise the furry growths that appear on stale bread and rotting fruits, and since time immemorial, fungi have been exploited for the production of leavened bread and alcoholic beverages. Indeed, so important are fungal products to humans that bread and wine are two of the principal symbols of the Christian Church, representing the body and blood of Christ. The role of fungi in the decay of vegetable and animal matter has also long been recognised. Fungi have been so intimately linked with the decomposition of organic matter that they have become synonymous with mouldiness, putrefaction and decay.

Fungi range dramatically in size from relatively large and compact structures such as puff-balls, mushrooms, toadstools and bracket fungi that can be seen attached to decaying trees, through the diverse network of filaments in soil, frequently associated with plant roots, down to the microscopic unicellular yeasts. However, all fungi are eukaryotic organisms. They possess a membrane-bound nucleus and nucleoli, cellular respiration occurs in mitochondria present in the cytoplasm, and fungal cells have an elaborate arrangement of internal membrane systems.

1.3.1 Moulds and their structures

The fungi display an astonishing variety of size and shape, but can be broadly divided into two groups, the **moulds** and the **yeasts**. The moulds are also referred to as filamentous or mycelial fungi, and they are composed of a network of filaments called **hyphae** (singular: hypha) that are interwoven into a structure called a **mycelium** (plural: mycelia). *Huphe* is the Greek word for a web, and mycelium is derived from *mukes* meaning mushroom. Moulds reproduce asexually or sexually, and the mycelium and **fruiting bodies** of a mould are collectively referred to as the fungal **thallus**, this being the Greek for green shoot. The spores of fungi are of primary importance in the identification of fungi, and are discussed in detail below. Mycelial tissue is also sometimes referred to as an **anastomosis** because it comprises a web of cross-connecting hyphae. The development of mycelial structures is by growth from the tips of the hyphae, with branching of the filaments occurring intermittently. The cytoplasm of young hyphae fills the filament, but further back from the growing tip, the cytoplasm becomes increasingly vacuolated, and the oldest hyphae are empty structures that may become cut off from the rest of the mycelium.

In the majority of moulds, hyphae are divided into sections by the regular occurrence of cross-walls or **septa** (singular: septum). These structures add rigidity to the filament, and help to control the flow of nutrients through the mycelial network. Septa vary in complexity; simple septa have a single central pore, but some septa seen in higher fungi have a **dolipore** structure, in which a narrow central pore is flanked by a cap-like, perforated membranes (parenthesomes) made up of amorphous material (Fig. 1.3). Individual compartments in the septate hypha may contain a single nucleus, and these are said to be uninucleate or they may contain several nuclei, and are thus described as multi-nucleate. In more primitive fungi such as the Phycomycetes, there are no septa to divide the hyphae into sections, and the aseptate hypha is described as coenocytic (Greek: *kinos*, common; *kutos*, a vessel). Septa do occasionally develop in Phycomycetes, but their function is to separate the reproductive structures from the vegetative body of the fungus, or to cut off the old sections of the thallus. Unlike the septa of other fungi, the structures elaborated by the Phycomycetes are solid plates, and they do not have a central pore (Fig. 1.4).

In certain higher fungi, adjacent hyphae can fuse vegetatively to give a three-dimensional network structure. It is from such structures that reproductive fruiting bodies are formed. Hyphae can also aggregate to give rise to other specialised structures that exhibit a high degree of internal organisation as a result of coordinated growth. **Rhizomorphs**, literally resembling root structures, are rope-like strands that have a highly differentiated structure. These structures appear to develop in response to stress, and in nature they develop in relatively dry environments such as are found in sandy soils. **Sclerotia** (singular: sclerotium) are hardened structures that enable certain moulds to survive in a dormant state. In culture, sclerotia are pigmented, and are sufficiently large to be seen by the naked eye. They are generally rounded, but may display an irregular shape. The cells of the outer wall of a sclerotium possess thick walls, and thus the structure has a thick, protective coat. This encloses a central cortex of hyphae that contains the food reserves necessary for dormancy. Nutrients are typically stored either as oil droplets or as glycogen. The most familiar structures generated by fungi, however, are the **mushrooms** and **toadstools**. These are highly complex reproductive structures that demonstrate an astonishing level of internal differentiation and organisation. A stalk or **stipe** supports the cap or **pileus** under which the **gills** develop. It is from these gills that spores are released. The development of gills is a highly coordinated process that responds to environmental stimuli. In order for

(a)

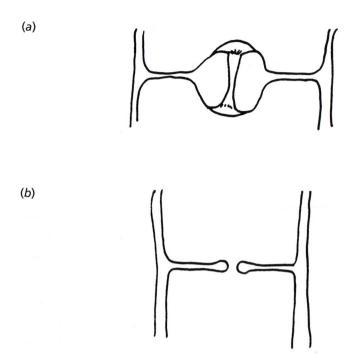

(b)

Fig. 1.3. (a) A dolipore and (b) a 'normal' septum.

(a)

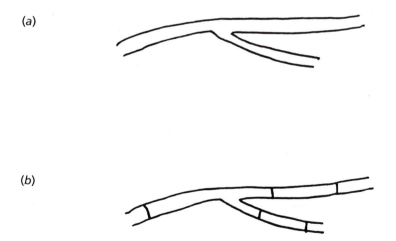

(b)

Fig. 1.4. Aseptate (coenocytic) and septate hyphae. (a) An aseptate or coenocytic hypha; the hypha shown in (b) is divided by regular septa.

spores to be released efficiently, gills must develop vertically. This is achieved by **geotropism**, as shown by the observation that if the developing structure is tilted, then the gills will still form to lie vertically.

1.3.2   Yeasts and their structures

In yeasts, the fungal thallus is generally a single cell. Yeasts are predominantly unicellular fungi that are round, oval or elongated in shape. They vary from 2 to about 10 micrometres in size. A limited number of yeasts elaborate extra-cellular capsules. An example is *Cryptococcus neoformans*, a human pathogen that causes a chronic form of meningitis increasingly seen in patients with acquired immune deficiency syndrome (**AIDS**). In this instance, the capsule mucopolysaccharide helps the yeast to evade the body's defence mechanisms, and thus allows it to cause disease.

Yeasts generally reproduce by the asexual process of **budding**. The parental cell develops a protruberance that swells and enlarges into a **blastospore** that eventually separates from its parent (Greek: *blastos*, sprout). However, in **fission yeasts**, such as *Schizosaccharomyces pombe*, a parental cell divides into two **progeny** in a manner somewhat similar to the transverse binary fission seen in bacterial reproduction. Yeasts rarely form true multicellular structures. Some yeasts form chains of elongated cells that are called **pseudomycelia** (singular: pseudomycelium) or **pseudohyphae** (singular: pseudohypha). Pseudomycelia are elongated yeast cells that arise from buds adhering together in branching chains. The individual cells within a pseudomycelium are independent of one another and, unlike the units within the septate hyphae of moulds, they are not connected by pores. Yeast cells with a typical unicellular morphology may cluster terminally or along the side of a pseudomycelium. These are called secondary blastospores. Some yeasts can produce septate, true mycelia under certain growth conditions (Fig. 1.5).

1.3.3   Dimorphic fungi

Although it is convenient to divide fungi into two groups, moulds and yeasts, there are fungi that are capable of adapting their structures in response to changes in their environment. They may grow in either a mycelial or a yeast form, depending on the prevalent growth conditions. These are referred to as **dimorphic fungi**. The mycelia formed by dimorphic fungi are true mycelia, unlike the pseudomycelia produced by some yeasts. Many of the fungi that

(a)

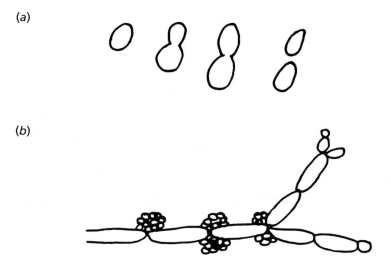

(b)

Fig. 1.5. Budding yeast cells (blastospores) and pseudomycelia that can develop in certain yeast species. The budding yeast cells are shown in (a) and the pseudomycelia are represented in (b). Elongated yeast cells are produced in chains, and secondary blastospores arise at cell junctions.

cause diseases in humans and animals are dimorphic, for example *Candida albicans*, the fungus that causes thrush. This is an infection that affects mucous membranes such as are found in the mouth or the genital tract. The visible symptoms of oral thrush are white plaques in the mouth, and vaginal thrush appears as an itchy white vaginal discharge.

1.3.4   The fungal cell wall

Fungi have been described as plant-like, because they are generally non-motile, and because their cells are bounded by a well-defined, multi-layered **cell wall**. However, the cell wall structure of plants and fungi differs considerably, with plant cell walls being made of celluloses and hemi-celluloses, and the fungal cell walls composed mainly of other polysaccharides including **chitin**, a polymer of **N-acetylglucosamine**. As well as being one of the principal components of fungal cell walls, chitin is a structural polymer that is found in the exoskeleton of the arthropod invertebrates.

The major component of the cell walls of both moulds and yeasts is **polysaccharide**, with up to 80% of the cell wall material comprising crystalline microfibrils in an amorphous matrix material. Of the remaining 20% of cell

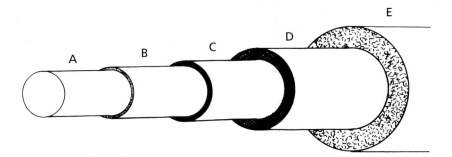

Fig. 1.6. The five-layered fungal cell wall structure. The fundamental layer, A, is the plasmalemma. Beyond this there is a layer, B, of chitin microfibrils in a matrix of proteins, mannans and glucans. This supports protein layer, C, beyond which is a glycoprotein network, D. The thickest, outermost layer, E, comprises amorphous glucans.

wall components, protein and lipid are present in approximately equal proportions. The cell wall polysaccharide component depends upon the type of fungus. In moulds, chitin is the principal fibrillar component and polymers of glucose known as **glucans** form the amorphous matrix material. In contrast to moulds, **mannans**, polymers of mannose, are the predominant structural components of yeast cell walls where they are also found with glucans. In baker's yeast, *Saccharomyces cerevisiae*, the cell wall contains less than 1% chitin, and this polymer is principally associated with bud scars, in which it forms a plug of material. In a special group of fungi (Oomycetes) within the Phycomycetes, cellulose forms the dominant structural component of the cell wall material.

The cell wall of fungi has five layers. It is exemplified by the architecture of the cell wall of the mature hyphae of the mould *Neurospora crassa*. The plasmalemma forms the foundation of the cell wall. Above this lies a layer of chitin microfibrils in an amorphous matrix of proteins, mannans and glucans that is about 20 nanometres thick. Beyond this lies a discrete protein layer of about 10 nanometres in thickness. This supports a **glycoprotein** network embedded in protein. This layer is about 50 nanometres thick. The outer layer of the cell wall is the thickest, and is made up of amorphous glucans. This layer is about 90 nanometres thick in *Neurospora crassa* (Fig. 1.6).

1.4 Reproduction in fungi

Reproduction in fungi may be **asexual** or **sexual** and, in both cases, **spores** are the structures that are responsible for dispersing progeny to colonise new locations. Some spores are designed to withstand adverse growth conditions or to provide for a period of dormancy. The mycelia of moulds may also become fragmented, and the resulting fragments may each subsequently develop into an individual thallus by the process of **vegetative reproduction**. The term vegetative reproduction is used to refer to asexual reproduction where special reproductive structures other than spores are not formed. The vegetative or asexual state of a fungus is known as the **anamorph** and the sexual state as the **teleomorph** (perfect state).

1.4.1   Asexual reproduction in fungi

The simplest form of asexual reproduction is the production of vegetative spores. There are two principal structures associated with vegetative reproduction. These are **arthroconidia**, and **chlamydoconidia**. *Arthron* is the Greek for joint, and arthroconidia are produced by the hyphae that become disjointed and fragment. These may also be referred to as **thallospores**. Chlamydoconidia are usually larger than arthroconidia, are rounder and are swollen with food reserves. The Greek word for cloak is *khlamus*, and chlamydoconidia derive their name from their thick cell wall. Formation of these structures is usually a response to environmental stresses. Under favourable conditions, both arthroconidia and chlamydoconidia germinate to produce new mycelia (Fig. 1.7(*a*)).

True asexual spores of fungi differ from vegetative spores in that they are formed on or in specialised structures called **sporophores**. They are also generally produced in large numbers. They arise because cells divide mitotically, and so the genetic composition of the spores is identical with that of the parent. There is an astonishing variety of size, shape, complexity and colour displayed by the asexual spores of fungi. This provides an excellent means of identifying fungi, and also in part forms the basis of mycological classification. Some fungi produce only one type of asexual spore, whereas others may produce different types of spore.

The majority of spores are disseminated by wind, water or insects. Asexual spores provide a means of reproduction for the fungus and, because of the large numbers of spores that are produced, widespread dissemination of the species is possible. Some asexual spores are short-lived, and are sensitive to external stresses such as ultraviolet irradiation and desiccation. However, this is not always the case. Some spores, particularly those that are darkly pigmented or thick-walled, are resistant to such environmental pressures. Resistant spores can be used as a dormant stage in the life-cycle of fungi.

(a)

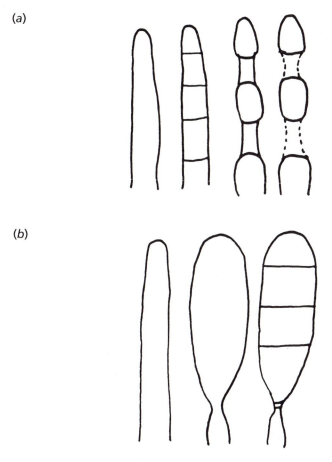

(b)

Fig. 1.7. Stylised arthroconidium (arthrospore) and thallospore production. (a) Production of arthroconidia, and (b) thallospore production.

Asexual spores can be produced either **exogenously** on the tips or sides of hyphae, or **endogenously** in specialised sac-like structures called **sporangia** (singular: sporangium). Lower fungi produce spores in sporangia that are typically formed at the tips of ordinary hyphae or on specialised hyphae called **sporangiophores**, although they may be formed along the hyphal structure. Sporangiophores usually contain large numbers of **sporangiospores** that may be motile or non-motile. Some sporangiophores form a **columella**. This is a dome-shaped structure that separates the sporulating and non-sporulating region within the sporangium. A columella of a sporangium that has discharged its spores may take on the appearance of an umbrella. Many terrestrial fungi produce non-motile spores called **aplanospores** (Greek: *a-*, not; *planao*, wandering), but aquatic and some soil-dwelling fungi that live in wet soils produce motile spores, called **zoospores**, within sporangia. The motility of zoospores is conferred by the possession of flagella. Some zoospores carry a single flagellum, whereas others typically carry a pair of flagella. In fungi that produce zoospores, the sporangium is referred to as the **zoosporangium**. Some mycologists classify these fungi in the **Protoctista** along with protozoa and nucleated algae.

Exogenously produced spores are often formed on specialised hyphae, and are referred to as **conidiospores** or more simply as **conidia** (singular: conidium; Greek: *konis*, dust). Conidia vary in shape, colour and complexity and may be large or small. In fungi producing two types of spores, the former are called **macroconidia** and the latter **microconidia**. They are borne on structures called **conidiophores** (Greek: *phero-*, to bear). Conidia are often produced from special narrow, pointed structures called **sterigmata** (singular: sterigma) that are attached to the conidiophore. Conidiophores sometimes, but not always, differ from vegetative hyphae, and they may be characteristic of a particular fungal genus or species. Conidiophores can become cemented together to produce an aerial stalk called a **coremium** or **synnema**. This raises the spore mass above the surface of the substrate, and so aids spore dispersal. In some fungi, conidiophores are organised inside a globose or flask-shaped structure called a **pycnidium**. Others, particularly plant pathogens, produce conidiophores in a flat or saucer-shaped bed called an **acervulus**, with spores discharged through an aperture called an **ostiole** (Fig. 1.8).

There are two primary ways in which conidia develop in fungi: blastic and thallic development. **Blastic conidia** arise by budding from or swelling of the hyphal structure from which they are separated by a septum (Fig. 1.7(*b*)), whereas **thallic conidia** are formed directly from a hypha by fragmentation or separation. Although arising by fragmentation of a hypha, thallic conidia may also undergo swelling. This is a characteristic of species of the genus *Microsporum*, one of the causes of ringworm in man, in which the hypha does not fragment during spore formation. Because the formation of conidia in these fungi involves the whole of the end of a hypha, they are described as **holothallic**, implying involvement of the complete fungal body. Thallic conidia may also be formed by fragmentation of the thallus into a series of segments and spore formation in this manner may be described as **arthric**. Fungi of the genus *Geotrichium* form conidia by fragmentation of a complete section of a hypha, and are thus described as **holoarthric**. The conidia of *Coccidioides* spp. are not formed in all sections of the hypha and are produced from alternate cells; these are described as **enteroarthric**.

Yeasts such as *Saccharomyces cerevisiae* that display asexual budding provide an

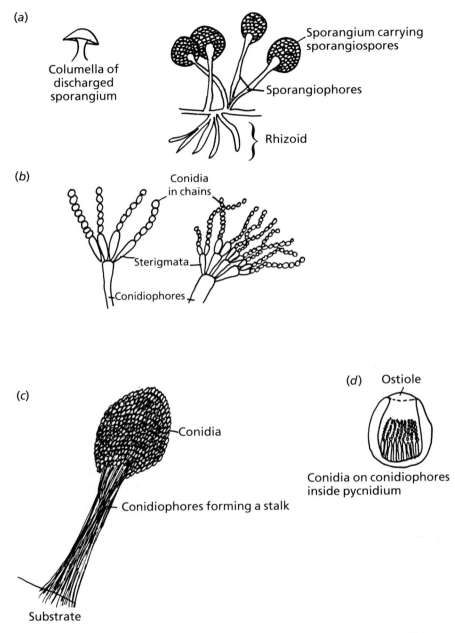

(a)

Columella of
discharged
sporangium

Sporangium carrying
sporangiospores

Sporangiophores

Rhizoid

(b)

Conidia
in chains

Sterigmata

Conidiophores

(c)

Conidia

Conidiophores forming a stalk

Substrate

(d)   Ostiole

Conidia on conidiophores
inside pycnidium

Fig. 1.8.   Some asexual structures produced by fungi. (a) Sporangia produced by
fungi of the genus *Rhizopus*. These fungi produce a columella which separates the
spores from the sporophore. In (b), the sterigmata and conidia produced by two
species of the genus *Penicillium* are illustrated. (c) A coremium, such as that produced
by *Stysanus* sp. and (d) a pycnidium as produced by *Coniothyrium* sp.

obvious example of blastic fungi. Although budding is the principal method of reproduction of yeasts, moulds may also form asexual spores in this manner. Fungi of the genus *Cladosporium* form chains of spores by repeated budding from a conidiogenous cell or fertile hypha. Similar patterns of spore development are seen with fungi belonging to the genera *Sporothrix* and *Alternaria*. These are all referred to as **holoblastic**. Alternatively, spores can swell or bud from within a hypha, and these are described as enteroblastic. This type of development is seen in members of two genera, *Scopulariopsis* and *Aspergillus*. Thus, within the two principal methods of spore formation, there are considerable variations on the basic themes, and the method of asexual spore development can be successfully exploited in the identification of fungi.

1.4.2   Sexual reproduction in fungi

Fungi are classified according to their method of sexual reproduction, but there is a group of fungi in which sexual reproduction has not been observed and they exist only in the anamorphic state. These fungi are collected together in an artificial group known as the **Fungi Imperfecti** or **Deuteromycetes** (**Deuteromycotina**). In some cases, Fungi Imperfecti may be incapable of sexual reproduction, but in other instances it may simply not yet have been observed. Some members of the Fungi Imperfecti are clearly related to species whose pattern of sexual reproduction has been elucidated. Such relationships are generally demonstrated by similarity in the asexual reproductive structures of the fungi. Occasionally it has been possible to remove species from the Fungi Imperfecti and to assign them a proper classification once sexual reproduction has been observed and their teleomorphic state characterised.

Sexual reproduction in fungi is often in response to environmental stresses such as temperature changes, adverse pH or nutrient depletion. In the laboratory, conditions can be manipulated to trigger sexual reproduction by growing cultures on media that are deliberately low in nutrients. The result of sexual reproduction in fungi is the production of sexual spores that are often resistant structures, capable of entering a dormant phase.

As with all sexual reproduction, that of fungi basically involves the fusion of two compatible nuclei. With the exception of some species of yeast, fungi exist in a **haploid** state. This means that they have a single set of unpaired chromosomes. Sexual reproduction in fungi thus produces a **diploid** state, where the chromosomes are paired, and cell fusion is followed by meiosis of the zygote nucleus (often immediately) so that the progeny can return once more to the haploid state. Sexual reproduction offers the fungi an opportunity to undergo genetic recombination to produce progeny that have a different genetic constitution from their parents. This has the advantage of fostering genetic diversity. In turn, this enhances the evolutionary possibilities that fungi can exploit through natural selection.

Typically there are three phases in sexual reproduction of fungi, namely **plasmogamy**, **karyogamy** and **meiosis**. Plasmogamy involves a fusion of two **protoplasts**. This brings together two compatible nuclei within the same cell. The pair of nuclei is called a **dikaryon** and the cell that contains them is described as **dikaryotic**. In the lower fungi, plasmogamy is almost immediately followed by karyogamy, or the fusion of the two nuclei, but in the higher fungi these two processes may be separated in time. Furthermore, in the higher fungi, dikaryotic cells may multiply, with a simultaneous division of the two nuclei in each cell. This process may yield numerous dikaryotic cells. This is referred to as the dikaryotic phase. When nuclear fusion or karyogamy does eventually occur, it is followed by meiosis, returning the fungal cells to a haploid state once more.

The sex organs of fungi are called **gametangia** (singular: gametangium). These are differentiated from vegetative hyphae. In some fungi they are indistinguishable from one another, whereas in others 'male' and 'female' gametangia are clearly different. Some fungi are capable of self-fertilisation, and are referred to as **homothallic** fungi. In contrast, **heterothallic** fungi require the interaction between two mating types, arbitrarily designated as e.g. + and −, for successful sexual reproduction to occur. The mode of sexual reproduction differs in the various classes of fungi.

Sexual reproduction in Phycomycetes   In all **Phycomycetes**, the result of sexual reproduction is a resting spore that germinates under favourable conditions to produce the asexual reproductive stage either directly, or shortly after germination. Aquatic Phycomycetes generally form motile **gametes** called **zoospores**. These fuse to form a motile **zygote** that only later enters a resting phase. Some Phycomycetes have morphologically dissimilar gametangia. The male structures are called **antheridia** (singular: antheridium), and the female structures are **oogonia** (singular: oogonium). The zygote that is produced from an antheridium and oogonium is called an **oospore**. In the more advanced Phycomycetes (**Zygomycota**), the gametangia are modified hyphae that are morphologically similar. Their fusion results in the production of thick-walled, resistant structures called **zygospores** (Fig. 1.9).

Sexual reproduction in Ascomycetes (Ascomycotina)   Sexual reproduction in the **Ascomycetes** results in the production of haploid sexual

ascospores inside a sac-like structure called an **ascus** (plural: asci). The shape of the ascospores and ascus varies with the particular species of fungus. There are typically eight ascospores in each ascus. In the lower Ascomycetes, including yeasts, the process of sexual reproduction is simple. Two vegetative cells fuse, and this is followed immediately by fusion of the two nuclei. The resultant zygote cell becomes the ascus in which meoisis occurs to produce the ascospores.

In the higher Ascomycetes, sexual reproduction is a more complicated process. The compatible nuclei are frequently to be found on morphologically dissimilar gametangia. The male gametangium is called an **antheridium** (plural: antheridia), and the female is an **ascogonium** (plural: ascogonia). Nuclear fusion, or karyogamy, does not immediately follow plasmogamy, or protoplast fusion, and dikaryotic cells are produced. These are also referred to as **ascogenous hyphae**. The asci that contain ascospores are produced from the terminal binucleate cells of ascogenous hyphae, and it is here that karyogamy and meiosis eventually take place to produce the ascospores. There are several types of asci. Some are globose to club-shaped releasing their ascospores on rupture of the wall, whereas others are cylindrical and have spore-ejecting mechanisms.

The higher Ascomycetes produce their asci in a fruiting structure or **ascocarp (Ascoma)**. There are three main types of ascocarp, namely a **cleistothecium**, a **perithecium** and an **apothecium** (Fig. 1.9). The cleistothecium is a completely closed structure which encloses the asci that lie randomly inside (Greek: *kleistos*, closed); ascospores can only be released by rupture or disintegration of the wall. The perithecium has an organised layer of asci on its inner wall, and an ostiole through which the ascospores are released. In many species of Ascomycetes that produce perithecia, each ascus moves up into the ostiole in succession, and there each forcibly discharges its ascospores. An apothecium is an open structure with asci arranged in a well-defined layer on its upper surface. Cleistothecia and perithecia are produced in culture, and are often visible to the naked eye.

Sexual reproduction in Basidiomycetes (Basidiomycotina)   The characteristic sexual cell in this group of fungi is the **basidium** (plural: basidia; Greek: *basidion*, basis). The basidium bears the haploid, sexual **basidiospores**, usually four, externally on structures called **sterigmata** (singular: sterigma; Greek: *sterigma*, support). In lower Basidiomycetes the basidium is septate and arises from a hypha or a thick-walled resting spore called a **teliospore**. In higher Basidiomycetes the basidia are unicellular and club-shaped and are usually produced on or in conspicuous fruiting structures

(a)

(b)

(c)

(d)

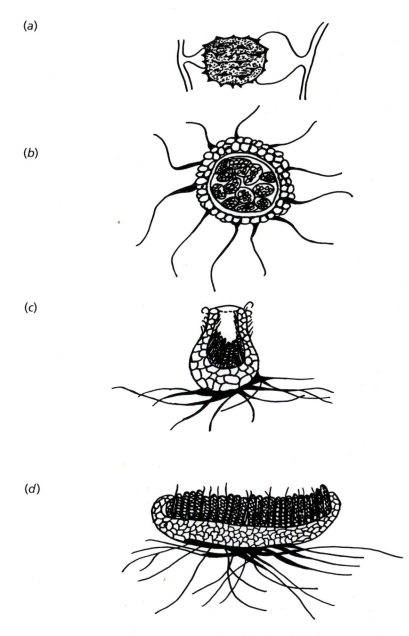

Fig. 1.9. Some sexual structures produced by fungi. (a) A zygospore between two suspensor cells of a fungus of the genus *Zygorrhynchus*; a Phycomycete. (b) A section through a cleistothecium, (c) a section through a perithecium, and (d) an apothecium showing asci with their ascospores. (b) to (d) all illustrate structures elaborated by Ascomycete fungi.

called **basidiocarps**. Often they are organised in a well-defined layer called a **hymenium** (Greek: *hymen*, thin skin or membrane) for example, on the surface of gills as in mushrooms and toadstools, or lining the pores of bracket fungi.

An extensive binucleate (dikaryotic) mycelial phase is characteristic of this group of fungi. It results from the fusion of adjacent hyphae without fusion of their nuclei; these must be from two sexually compatible strains if the fungus is heterothallic. A process called **clamp formation** ensures that when these dikaryotic cells multiply, and their two nuclei divide simultaneously, one of each pair goes into each daughter cell. The mycelia of many Basidiomycetes are therefore characterised by small bumps along the length of the mycelium, where the clamp-connections occur at the junction of the cells.

1.5   Nutrition of fungi

Fungi are often erroneously said to resemble plants. However, plants can elaborate complex organic compounds from simple inorganic molecules such as water and carbon dioxide by the process of photosynthesis. Fungi, in contrast, all require a supply of preformed organic compounds for their energy production and growth. They are thus described as **heterotrophic** organisms (Greek: *heteros*, other; *trophikos*, nourishment), hence heterotrophs are nourished from elsewhere, rather than being able to feed themselves as the **autotrophic** plants do. Most fungi are found in dark, moist habitats, but are universally present where organic matter is found.

Fungi may be **saprophytic** or **parasitic**. Saprophytic organisms are defined as those that live on decaying organic matter, and the term is derived from two Greek words, *sapros* and *phuton*, meaning decayed plant. Parasites (Greek: *para*, other or beyond; *sitos*, food) derive their nutrients from living plants or animals, and generally cause disease in their hosts. The majority of fungi are found as saprophytes in the soil, living on decaying plant material, where they play a vital role in the recycling of organic matter. Fungi feed by secreting hydrolytic enzymes into their local environment. These enzymes digest the various polymers to produce soluble products that the fungus then absorbs.

In artificial culture, many fungi can grow on a mineral salts medium containing a source of nitrogen salts, providing that they have glucose present as a carbon source. They can thus manufacture all the complex organic molecules that they require for growth from the metabolism of glucose. Other fungi require an exogenous supply of **vitamins** or other growth factors that they cannot manufacture themselves in order to grow in artificial culture. Added growth factors are of particular importance if vegetative cells in culture are to initiate sporulation. Fungi have a specific requirement for trace elements including calcium, magnesium, iron, zinc, copper and manganese. No fungus has been found that can fix atmospheric nitrogen.

Glycogen is the principal storage polymer used by fungi, but oil droplets are also used for nutrient storage. Like most eukaryotes, fungi are **obligate aerobes**, and they obtain energy from the aerobic respiration of glucose. However, a minority of yeast species are **facultative anaerobes**, and if held under anaerobic conditions, they can obtain energy by **fermentation.** This process is not as efficient as respiration. In culture, mutants of the baker's yeast *Saccharomyces cerevisiae* that lack mitochondria occur at a frequency of about 1%. These cells are incapable of respiration, because of their inability to assemble mitochondria. Consequently, when they grow on a glucose-based solid medium, they can obtain energy only by fermentation. Such mutants give rise to very small colonies when compared to the **wild-type** cells that do possess mitochondria, and are thus described as **petite** mutants.

1.6   Bacterial morphology

Prokaryotic cells are very small. On average, bacteria range in size from between 0.2 micrometres and 10 micrometres, although spiral organisms may extend up to 100 micrometres in length. The majority of bacteria are less than 5 micrometres long. One important consequence of this is that prokaryotes have very high ratios of surface area to volume. This is reflected in a relatively large surface in contact with the environment, and over which metabolites may enter and leave the cell. Within the cell, there is a relatively small volume through which metabolites have to travel. Bacterial cells may, therefore, dispense with mechanisms for the distribution of nutrients within cells, such as the cytoplasmic streaming seen in eukaryotic cells, and rely instead on passive diffusion. This, however, places severe physical limitations upon the size of prokaryotic cells.

The Greek word *bakterion*, from which **bacterium** (plural: bacteria) is derived, means a small stick. This implies that bacteria are stick-shaped. In fact, bacterial cells are of three basic shapes: they are either round, rod-shaped or spiral in form. Although bacteria may be divided into just three basic shapes, there is an astonishing variety of forms seen within these morphological groups, and some degree of variability on these patterns may be discerned (Fig. 1.10). The characteristic shape and staining properties of bacteria

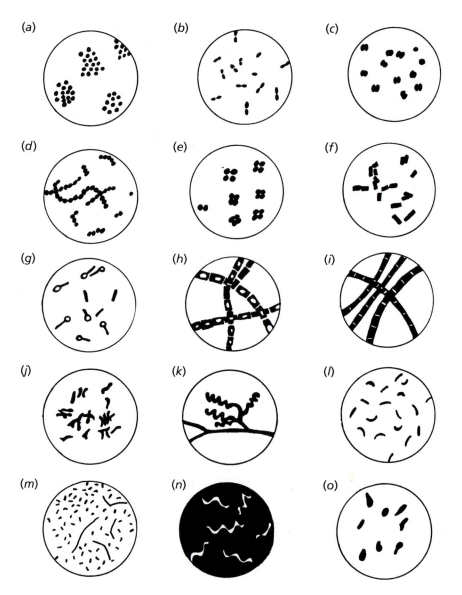

Fig. 1.10. The microscopic appearance of representative bacteria.
(a) *Staphylococcus aureus*; (b) *Streptococcus pneumoniae*; (c) *Neisseria gonorrhoeae*;
(d) *Streptococcus pyogenes*; (e) *Sarcina* sp. (f) *Clostridium perfringens*; (g) *Clostridium tetani*; (h) *Bacillus anthracis*; (i) *Beggiatoa* sp. (j) *Corynebacterium diphtheriae*;
(k) *Streptomyces viridochromogenes* (viewed under low power); (l) *Vibrio cholerae*;
(m) *Haemophilus influenzae*; (n) *Borrelia recurrentis* (viewed using dark-ground microscopy); (o) Bacteroids of *Rhizobium trifolii*.

present in clinical samples may, in certain circumstances, permit a provisional identification to be made simply from an examination of a Gram-stained film. Clusters of rounded bacteria that appear blue-black when stained using Gram's method (see below) if seen in **pus** aspirated from a bone in suspected cases of **osteomyelitis** is sufficient to permit the provisional identification of *Staphylococcus aureus* as the cause of the infection. For many years, bacterial morphology was the foundation stone of bacterial **taxonomy**, but with the application of genetic and molecular biological techniques to this problem, the shape that bacteria assume is becoming less important.

Bacteria that have an approximately round shape are referred to as **cocci** (singular: coccus). This name is derived from a Latinised Greek word *kokkos*, meaning a berry. Cocci are not necessarily perfectly spherical, and may be quite markedly deformed. Cells of *Streptococcus pneumoniae* generally occur in pairs, known as **diplococci**, that are flattened to appear like small lancets. For this reason they are sometimes referred to as **lanceolate diplococci**. Another diplococcus, *Neisseria gonorrhoeae*, has cells that appear flattened against one another. The cells of *Streptococcus pneumoniae* are aligned on their long axis, with their short axes in parallel, whereas those of *Neisseria gonorrhoeae* are aligned along their short axis, with their long axes in parallel. Cocci may also associate in clusters of more than two cells. Streptococci appear in chains as a result of regular cell divisions in one plane (Greek: *streptos*, twisted). Staphylococci bunch like grapes as a result of cell division along irregular planes (Greek: *staphule*, grapes). Bacteria may grow in cuboidal packets as a result of regular cell divisions along three planes. An example of this form of growth is afforded by bacteria of the genus *Sarcina*.

The term **bacillus** (plural: bacilli) is derived from the Latin word *bacillus*, meaning a stick. Bacilli are rod-shaped and come in a variety of forms. Even within a single bacterial genus, considerable variation in shape is seen. For example, cells of *Clostridium perfringens* are relatively short and fat, and the vegetative cell shape is not deformed by the presence of a spore, whereas cells of *Clostridium tetani* are long and slender. Often the end of the cell is swollen to accommodate its spore. This gives the cells of *Clostridium tetani* the appearance of drumsticks.

Just as many cocci arrange themselves in specifically shaped groups, some bacilli are associated with particular arrangements. *Bacillus anthracis*, the causative agent of anthrax, grows as a long chain of cells aligned along the long axis. Cells of the genus *Beggiatoa*, and many cyanobacteria form chains in which the cells are in intimate contact. Cells align in long strands in which individual cells are compartmentalised by cross-walls. Such chains are called **trichomes** because they resemble hairs (Greek: *trikhos*, hair). Bacteria of the genus *Corynebacterium* have cells that align along their long axis in parallel, in a manner somewhat similar to fence posts. This arrangement is referred to as a **palisade**, the Latin word for a stake being *palus*. Some microbiologists describe the arrangement of corynebacteria as resembling Chinese letters.

Streptomycetes form long, multinucleate hyphae, that become branched. A collection of hyphae is called a mycelium. The terms hyphae and mycelia are also applied to analogous fungal structures.

When viewed in the light microscope, some bacteria have a rod shape that is so truncated that their cells appear almost round. Electron microscopy does make the shape of such cells easier to distinguish. These cells are referred to as **coccobacilli** because of the difficulty in determining their nature.

Curved rods are referred to as **vibrios**. This name reflects the vibrational motility seen among these bacteria. Bacterial cells may grow in a twisted, helical shape. Spirilla have rigid cells whereas spirochaetes have a very flexible structure. The Greek word for a coil is *speira*, Latinised as *spira*, and *khaite* is Greek for long hair.

Pleomorphic bacteria have variable shapes (Greek: *pleon*, more; *morphe*, shape). Many bacteria display a degree of pleomorphism. In cultures of *Proteus mirabilis* cells occasionally grow as long filaments, whereas the majority of cells grow as short rods. Similarly *Haemophilus influenzae* cells may grow as filaments, rods or even as coccobacilli. Corynebacteria, although they appear as bacilli are very irregular in shape. In some cases pleomorphism is the result of the environment in which the bacteria are grown. In artificial culture, bacteria of the genus *Rhizobium* grow as regularly shaped bacilli of fairly uniform dimensions, but when the same bacteria are seen in microscopic preparations of the root nodules of nitrogen fixing plants, cells have a highly degenerate and irregular appearance and are referred to as **bacteroids**.

1.7   Bacterial cell structure

As a consequence of their small size, bacterial cells can function without subcellular organelles. In spite of, or perhaps because of their size, bacteria are the only life-forms successfully to colonise some of the most hostile environments on Earth. Although the typical prokaryotic cell has a relatively simple structure, some bacteria have evolved subcellular structures that enable them to obtain the maximum benefit from their environment.

1.7.1 The nucleoid

In contrast to eukaryotic cells that have a variable number of chromosomes contained within a membrane-bound nucleus, bacteria have a single chromosome comprising a circular molecule of DNA. When linearised, this may measure up to 1 millimetre. The chromosomal material is not present in a membrane-bound nucleus, but, nevertheless, is located in a distinct region of the cell called the **nucleoid**. For much of its length, bacterial DNA is tightly packaged in a supercoiled form, similar to the structure adopted by a twisted elastic band. This structure becomes relaxed for the **transcription** process whereby messenger RNA is formed as a precursor to protein synthesis. Supercoiling is also relaxed during DNA replication. DNA gyrase is the enzyme responsible for maintaining the supercoiled DNA structure, and represents one of the proteins intimately associated with bacterial DNA. Enzymes involved in the transcription and replication of DNA, in DNA recombination and in the regulation of gene function are similarly found in association with prokaryotic DNA. However, bacteria lack **histones**: large, basic proteins attached to the DNA in eukaryotic nuclei.

The bacterial chromosome carries all the genetic material necessary for survival under normal conditions, but very many bacteria also carry **plasmids**. These are extrachromosomal DNA molecules that encode functions not normally required for growth. Plasmids may give their host bacterium a selective advantage in certain special circumstances. Most bacterial plasmids are circular in structure, but linear plasmids have been described in the genus *Streptomyces*. Plasmid sizes are most often expressed in kilobases (kb), one kilobase representing 1000 base-pairs. Plasmids range in size from about 0.5 kb up to several hundred kilobases, and one cell may carry several plasmids of various sizes. Plasmids are capable of autonomous replication and are stably inherited. Single copy plasmids are present as one copy per cell; multicopy plasmids may be represented by up to 60 DNA molecules per cell. Large plasmids tend to be of low copy number.

The most extensively studied plasmids are those encoding antibiotic resistance, but plasmids may also carry the genes encoding surface adhesion, toxin production, resistance to heavy metal ions, or the ability to utilise unusual metabolites. **Toxins** may cause disease in higher organisms, or may damage bacteria. Toxins that act against bacteria are called **bacteriocins**. Bacteriocins are highly specific in their toxicity, and will affect only a narrow spectrum of bacteria, generally closely related to the producer bacterium. It is thought that bacteriocin production has evolved as a means whereby individual strains obtain a selective advantage over close relatives in particular habitats. Bacteria that elaborate bacteriocins must also have a means of protecting themselves from their own toxic products, and the genes encoding bacteriocin immunity are also carried on plasmids. Although some plasmids may be identified with a particular function, many have no known role, and are called cryptic plasmids.

Some plasmids have the ability to insert themselves into the bacterial chromosome and, in appropriate circumstances, to excise themselves. The process of excision from the chromosome may be imprecise, and chromosomal genes may become associated with plasmids in this way. Conversely, plasmid genes may become permanently integrated into chromosomal DNA following the imprecise excision of a plasmid. This lends a degree of fluidity to the bacterial **genome**. Genetic fluidity is enhanced by the ability of some plasmids to transfer copies of themselves into another strain. This may occur not just within species, but often between genera and even between different bacterial families.

Genetic instability in bacteria is enhanced by the activities of **transposable elements**. These are sequences of DNA that are capable of jumping from one location to another, unrelated, site in the genome. Transposable elements may be located on the chromosome or on plasmids, and they can jump between these two types of DNA structure. They may also transpose from one plasmid into another. The site into which the transposable element jumps need not have any sequence identity with either the transposable element or its surrounding sequences. Consequently, **transposition** is sometimes called **illegitimate recombination**. Transposition is often a random process, with DNA elements inserting into new sites indiscriminately, showing no preference for the target sequence into which insertion occurs. Some transposable elements do show a regional preference for target sequences, and are more likely to insert into particular sequences. The random nature of the transposition process exhibited by many transposable elements has been exploited to produce insertional mutants. These are of great value in the study of bacterial genetics.

The simplest transposable elements, known as **insertion sequences**, simply encode their own transposition functions. **Transposons** are more complicated structures and may carry genes other than those required for transposition. These include genes encoding antibiotic and heavy metal ion resistance, use of unusual metabolites or toxin production. The first bacterial transposon to be discovered encodes resistance to ampicillin, an antibiotic used to treat infections caused by Gram-negative bacteria such as *Escherichia coli*. Mercury resistance is often associated with transposons. There are transposons that encode the ability to break down urea or to ferment lactose, and there is evidence that the ability of strains of pseudomonads to metabolise toluene is encoded by a transposon. *Escherichia coli* strains are associated with traveller's diarrhoea, and one of the toxins responsible is found on a transposon. Transposons may carry several genes not associated with transposition functions, and some transposons may carry genes whose functions are cryptic. Often, the mechanism of transposition involves replication of the transposable element. One copy remains at the donor site and a new element is inserted into the target site. Other transposable elements are excised from the donor site prior to recombination into their new location.

Replication of bacterial DNA is associated with the cell membrane. This applies to both chromosomal and plasmid DNA. The regions of association of different plasmids with the bacterial membrane vary, and if two plasmids have similar membrane attachments, they cannot co-exist in the same cell. They are said to be **incompatible**. Plasmids may be grouped by observing their incompatibility with other plasmids. This is referred to as incompatibility typing, and groups of plasmids defined in this way are called *inc* groups.

1.7.2   The bacterial cytoplasm and membrane

The cytoplasm of prokaryotic cells lacks membrane-bound organelles, and in many bacteria is relatively featureless. It is the site of protein synthesis. Bacterial messenger RNA is translated directly into **polypeptides** as it is transcribed, and does not require the processing undergone by eukaryotic messenger RNA. A DNA sequence encoding a gene in a eukaryotic cell carries additional sequences called **introns** that do not appear in the mature messenger RNA. These are sliced from the messenger RNA. The DNA sequences that encode the mature messenger RNA are known as **exons**. Many ribosomes are aligned along a messenger RNA molecule in the bacterial cytoplasm. Each ribosome has its **nascent peptide** attached. Such aggregations are known as **polysomes**.

Bacterial cytoplasm may also contain **inclusion bodies** that may be organic or inorganic in nature. Organic inclusion bodies usually contain poly-β-hydroxybutyrate or glycogen. These act as energy storage reservoirs. **Volutin granules** are characteristic of the corynebacteria. These are composed of polyphosphate molecules and are also known as **metachromatic granules** because they stain a reddish colour with methylene blue rather than the blue expected with this dye. Volutin granules act to store phosphates. **Sulphur granules** comprise the second type of inorganic inclusion body found in the bacterial cytoplasm. They are found in bacteria that are able to oxidise hydrogen sulphide to produce elemental sulphur.

The prokaryotic cell membrane may become invaginated to form **vesicles**, tubules or **lamellae**. The membrane may also become invaginated to form structures known as **mesosomes**. The exact nature of these structures has been the cause of considerable controversy for many years. The presence of these structures has been attributed to artefacts created during sample preparation for electron microscopy, and in some instances this is almost certainly the case. However, bacteria that have atypical lifestyles often have an increased requirement for membranes with a larger surface area than is normally seen. This is to enable them to carry out the numerous metabolic functions that require membranes and that help them to survive in unusual conditions. There is, therefore, little doubt that membrane invaginations do represent real structures in particular bacteria. If located in the middle of cells, mesosomes are referred to as central mesosomes; those at the edge of the cell are peripheral mesosomes. Several different functions have been ascribed to mesosomes. They probably play a part in cell division, possibly assisting chromosomal segregation, ensuring that each daughter cell receives a copy of the chromosome. Mesosomes have a very large surface area and are thought to be involved in the export of proteins. Cellular **respiration** and photosynthesis are biochemical processes requiring membranes, and these processes are also associated with mesosome structures. In bacteria that have a very high rate of respiration there are extensive internal membranes to support the enzymes of the electron transport chain. Similarly, photosynthetic bacteria may have internal membranes in **chromatophores** or thylakoids.

Aquatic bacteria may have **gas vacuoles** to aid buoyancy. Gas vacuoles are found particularly in photosynthetic bacteria and help to maintain cells at the top of water. Generally, light does not penetrate very far into a body of water, so it is important for photosynthetic organisms that they live as close as possible to a light source. Although gas vacuoles are permeable to gas they are impermeable to water. They collapse under pressure, and the bacteria in which they are found then sink.

1.7.3   The bacterial cell wall

Although there are a few bacteria such as the mycoplasmas and certain archaebacteria that lack cell walls, the vast majority of bacteria do elaborate such structures. The **plasma membrane** and the cell wall together comprise the cell envelope. The bacterial cell wall has evolved to provide its cell with a rigid cage that lends it a characteristic shape. It also provides a strong coat to protect the cell contents from external stresses.

Eubacteria can be divided into two groups depending upon their ability or

(a)

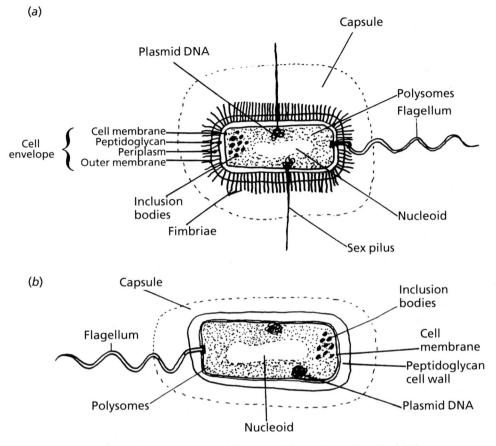

Fig. 1.11. Stylised Gram-negative and Gram-positive prokaryotic cells. (a) The structures associated with Gram-negative bacteria, and (b) the structures of Gram-positive bacteria.

inability to retain a crystal violet–iodine complex when exposed to an organic solvent such as acetone or alcohol. Those bacteria that retain the complex are called **Gram-positive**; those that cannot are said to be **Gram-negative**, so called after Christian Gram who described this stain in 1884. The precise mechanism of the Gram stain is not fully understood, but the Gram reaction is intimately related to the structure of the cell wall. Gram-positive bacteria have cell walls that are fundamentally different from those of Gram-negative bacteria. It is noteworthy that although archaebacteria have cell walls of a different structure from those of the eubacteria, some do have the ability to retain the Gram stain and so appear Gram-positive, whilst others cannot and appear Gram-negative (Fig. 1.11).

The Gram-positive cell wall   The principal component of the Gram-positive cell wall is **peptidoglycan**. This polymer has other names, and is also known as **mucopeptide** or **murein**. It is a uniquely bacterial polymer that is also exceptional in being the only biological polymer containing both L- and D-amino acid isomers. The detailed composition of peptidoglycan varies from species to species, but the fundamental structure is the same in all eubacteria with cell walls. Peptidoglycan comprises a disaccharide polymer joined by a peptide bridge attached to one of the sugars. The disaccharide comprises *N*-acetyleglucosamine joined to **N-acetylmuramic acid**. A tetrapeptide is joined to the *N*-acetylmuramic acid. The tetrapeptide contains both L- and D-amino acids, and the terminal amino acid is almost always D-alanine. The tetrapeptide chains are cross-linked with oligopeptide bridges. In *Staphylococcus aureus* the cross-bridge is composed of five L-glycine residues. The polysaccharide component of peptidoglycan is constant; it is the oligopeptides that give the structure its species variety. Despite this variability, nearly all bacteria have an L-alanine residue attached to the *N*-acetylmuramic acid and a D-glutamine residue joined to the L-alanine. The cross-linking of chains seen in peptidoglycan bestows great strength upon this structure (Fig. 1.12).

Gram-positive cell walls have up to 40 layers of peptidoglycan in their cell wall. Consequently, a very large proportion of the cell's metabolism is devoted to the production of cell wall material. This gives an indication of the importance of the cell wall to Gram-positive bacteria, and these structures can withstand enormous stresses. It is said that the cell wall of a typical Gram-positive bacterium may withstand pressures of up to 200 atmospheres (1 atm = 101325 Pa). Because of interlinking between layers, the entire cell wall structure may represent a single molecule of peptidoglycan. This lends the wall rigidity and strength as well as determining the shape of the bacterium.

The peptidoglycan of Gram-positive bacterial cell walls is supplemented by other polymers, principally **teichoic** and **teichuronic acids**. Together these may constitute up to half the cell wall structure of Gram-positive cell walls, although in other cases up to 90% of the cell wall is composed of peptidoglycan. The supplementary polymers of Gram-positive cell walls are covalently linked to the peptidoglycan structure, and may also be linked to the bacterial membrane, providing anchorage of the cell wall to the cell membrane. Teichoic and teichuronic acids constitute the major surface **antigens** of Gram-positive bacteria.

The Gram-negative cell wall   The structure of the Gram-negative cell wall is more complex than that of Gram-positive bacteria. The plasma membrane forms the inner membrane of the cell envelope. Beyond the plasma

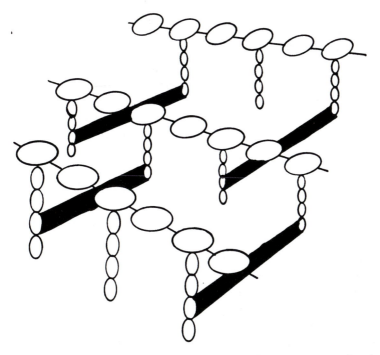

Fig. 1.12.   Part of a peptidoglycan sheet from *Staphylococcus aureus*. The glucan backbone, shown as chains of large ellipses, comprise alternating units of *N*-acetylglucosamine and *N*-acetylmuramic acid. The primary tetrapeptide chains are attached at one end to *N*-acetylmuramic acid residues. In *Staphylococcus aureus*, the tetrapeptide is made up of L-alanine, D-isoglutamic acid, L-lysine and D-alanine. A cross-bridge links the L-lysine and D-alanine residues on the primary tetrapeptide chains, and this is composed of 5 glycine residues. Peptidoglycans from different bacteria have a different detailed structure, but all possess the glucan backbone and primary amino acid residue chain. For example in *Escherichia coli*, the L-lysine residue found in *Staphylococcus aureus* is replaced with di-aminopimelic acid, and this links directly with a D-alanine residue on another tetrapeptide chain, without the pentaglycine bridge.

membrane lies the **periplasmic space**, bounded on the outside by a thin layer of peptidoglycan to which is attached an **outer membrane**. The term periplasmic space implies an empty area. In fact this is not the case. The periplasmic space provides a buffer between the inside of the cell and its external environment. It is a region of the cell in which the microenvironment may be easily controlled, and it is a site of considerable metabolic activity, particularly concerned with the transport of metabolites. The space is packed so tightly with enzymes, proteins, metabolites, etc. that it has the consistency of a gel. All bacteria export proteins into their environment. Gram-positive bacteria are continually losing proteins into their surroundings, but in Gram-negative bacteria many exported proteins are retained within the periplasmic space (Fig. 1.13).

The peptidoglycan of Gram-negative cells has the same fundamental structure as that found in Gram-positive cells, but Gram-negative cells generally have only one or two layers. Peptidoglycan is linked to the Gram-negative outer membrane by lipoprotein bridges. **Lipoprotein** is the most abundant protein found in Gram-negative bacteria, and provides an anchor for the outer membrane. This membrane is a lipid bilayer structure in which the phosopholipids present in the outer layer of typical biological membranes are replaced by **lipopolysaccharides**. The outer membrane lipopolysaccharide of Gram-negative bacteria comprises a complex lipid, lipid A, attached to a core polysaccharide, which, in turn, is attached to a terminal repeating polysaccharide unit, projecting outwards. Lipopolysaccharides constitute up to 40% of the surface structure of Gram-negative bacteria, reflecting their importance (Fig. 1.14).

Lipopolysaccharide acts as the major somatic antigen of Gram-negative bacteria. Somatic antigens are known as 'O' antigens, from the German word *ohne*, meaning without. Lipopolysacchride is toxic to animals and humans, and is also known as **endotoxin**. Purified endotoxin can cause fever and the symptoms of **septic shock**. These include mental confusion, shaking chills, a rapid heart beat and low blood pressure. The loss of an adequate blood supply may lead to tissue and organ damage, which, in turn, may be fatal. It is imperative that genetically engineered products obtained from Gram-negative bacteria must be purified and free from the endotoxin component of the bacterial cells.

Punctuating the outer membrane of Gram-negative bacteria are various protein channels. Some are non-specific and permit a variety of solutes in the environment to gain access to the periplasmic space: others are specifically associated with the transport of particular molecules that cannot pass through the non-specific channels at all, or in sufficient quantity to permit normal cellular activity. These pores are made from proteins referred to as **porins** or **outer membrane proteins**.

The cell walls of acid-fast and coryneform bacteria

There exists a group of bacteria that do not readily take up the Gram-stain or other stains. It comprises bacteria of the genera *Mycobacterium* and *Nocardia*. The cell wall structure of these bacteria is quite different from that of other bacterial cell walls. These bacteria can be stained by treatment with hot, strong, carbol fuchsin using the Ziehl Neelsen method, and they resist decolorisation by mineral acids. Thus, they are referred to as **acid-fast** bacteria. Coryneform

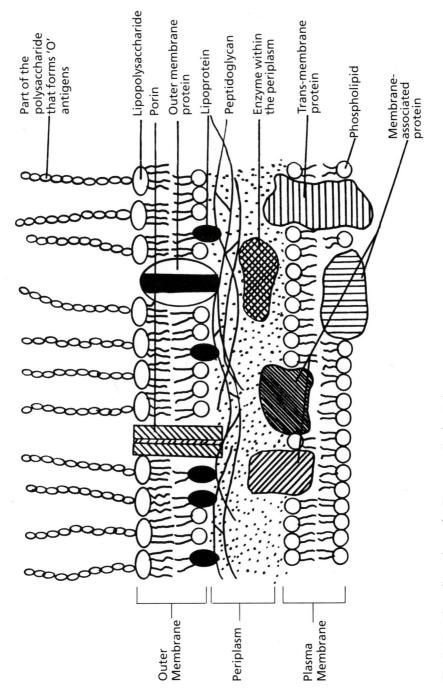

Part of the polysaccharide that forms 'O' antigens

Lipopolysaccharide

Porin

Outer membrane protein

Lipoprotein

Peptidoglycan

Enzyme within the periplasm

Trans-membrane protein

Phospholipid

Membrane-associated protein

Outer Membrane

Periplasm

Plasma Membrane

Fig. 1.13.   The cell envelope of a Gram-negative bacterium.

'O' antigen

Core
polysaccharide

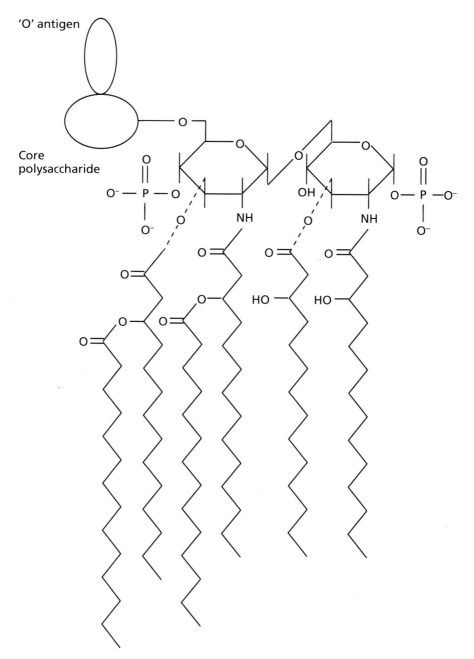

Fig. 1.14. Structure of the lipid A component of *Escherichia coli* lipopolysaccharide. The core polysaccharide and 'O' antigen are not drawn to scale.

bacteria, although not acid-fast, are genetically closely related to acid-fast bacteria, and they have a similar cell wall structure. Covalently attached to peptidoglycan are polymers whose subunits are composed of arabinose and galactose. The **arabinogalactan** structure is esterified to **mycolic acids**, complex fatty acids that give the cell wall a waxy nature.

The cell walls of archaebacteria   The archaebacteria comprise a large and heterogeneous group of bacteria. Representatives colonise some of the most extreme habitats on the planet. The structures of these bacteria have evolved to enable them to survive hostile conditions. Most archaebacteria elaborate cell walls. Unlike eubacteria, archaebacteria do not manufacture N-acetylmuramic acid, and as a consequence they cannot make peptidoglycan. Despite this, some archaebacteria appear Gram-positive when stained with Gram stain.

The plasma membranes of archaebacteria contain ether-linked branched aliphatic hydrocarbons. These are thought to confer greater mechanical strength upon membranes than do the ester-linked unbranched hydrocarbons typical of eubacterial membranes. The tougher membranes of archaebacteria probably aid survival in extreme conditions. **Thermoacidophiles** of the genus *Thermoplasma* can be isolated from coal slag-heaps and can live in an environment where the pH lies between 1 and 2 and where the temperature ranges from 55 to 59 °C. These bacteria lack cell walls, and rely upon the integrity of their plasma membranes to survive under such harsh conditions. Interestingly, thermoplasmas lyse at neutral pH. *Sulfolobus* is another thermoacidophile genus, but these bacteria have cell walls composed of protein.

Methanobacteria grow in organically rich habitats such as swamps and bogs. Bacteria belonging to the genera *Methanobrevibacter* and *Methanobacterium* appear as Gram-positive cells. Their cell walls contain **pseudomurein**. This polymer contains L-amino acids attached to polysaccharides. Pseudomurein also differs from peptidoglycan in that N-acetylmuramic acid is replaced by **N-acetyltalosaminuronic acid**, and the terminal D-alanine is replaced with L-glutamic acid. Other methanobacteria have cell walls comprising protein, but bacteria belonging to the genus *Methanosarcina* have cell walls containing heteropolysaccharide.

Extreme **halophiles** live in salt pans and in salt lakes such as the Dead Sea. The cell walls of bacteria of the genus *Halobacterium* are made up of proteins and they require a high salt concentration to maintain their integrity. At low salt concentrations, halobacteria lyse because their cell walls disintegrate. In contrast, members of the genus *Halococcus* have cell walls made of heteropolysaccharide, and these walls are stable even at low salt concentrations (Fig. 1.15).

(a)

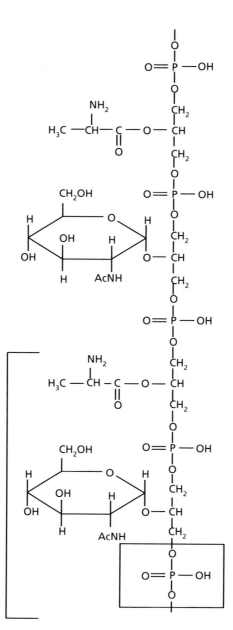

Fig. 1.15. (a) For caption see p. 40.

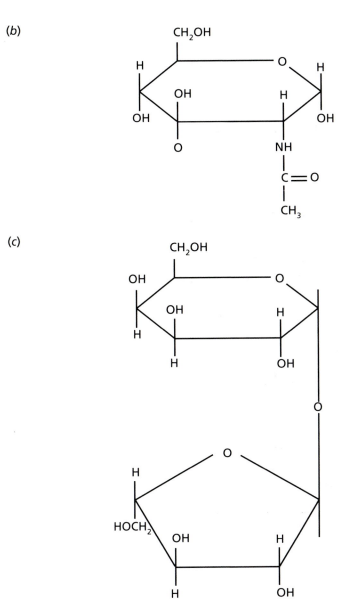

Fig. 1.15.   (b) and (c) For caption see p. 40.

(*d*)

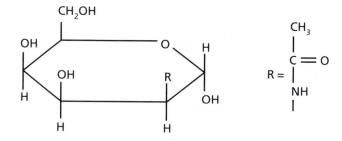

Fig. 1.15.   Some molecules associated with the cell walls of bacteria. (*a*) Two teichoic acid units with alternating side-chains of *N*-acetylglucosamine and alanine. In teichuronic acids, made during phosphate limitation, phosphoric acid residues (boxed in the figure) are replaced by sugar acids such as D-glucosuronic acid or *N*-acetylmannosuronic acid. (*b*) A molecule of *N*-acetylglucosamine. *N*-acetylmuramic acid is derived from *N*-acetylglucosamine by the addition of lactic acid to the third carbon atom through an ether bond. (*c*) A monomeric unit of arabinogalactan, a molecule associated with the cell walls of coryneform bacteria and mycobacteria. (*d*) *N*-acetyltalosamine – a component of the cell walls of certain archaebacteria.

Spheroplasts and protoplasts   Bacteria may lose peptidoglycan either as a result of maturation, or by growth in an **isotonic medium** in the presence of a *β*-lactam antibiotic such as penicillin. In Gram-positive bacteria this leads to the formation of **protoplasts**. Gram-negative cells lacking cell walls are called **spheroplasts**. Spheroplasts are differentiated from protoplasts because the former retain an outer membrane structure. Naturally occurring bacteria that lack peptidoglycan are referred to as **L-forms**, so called because they were first described in the Lister Institute. L-forms are pleomorphic, illustrating the importance of peptidoglycan in determing cell shape. If all traces of peptidoglycan are removed from spheroplasts or protoplasts, cell wall structures cannot regenerate. This demonstrates that bacteria lack the ability to generate peptidoglycan *de novo* but, rather, simply keep adding subunits to a pre-existing structure.

1.7.4   Structures external to the bacterial cell wall

Many bacteria elaborate structures beyond the cell wall. These structures have a variety of functions. They may help to prevent desiccation or nutrient loss, aid adherence to surfaces, prevent the action of **opsonising antibodies** and delay **phagocytosis** by white blood cells. In intracellular bacteria these structures may help to prevent killing inside phagosomes. They may also act as receptor sites for bacteriophages, viruses that infect bacteria. Mostly these

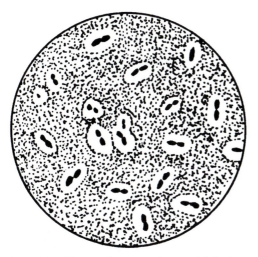

Fig. 1.16.   The quellung reaction. In this instance, a strain of *Streptococcus pneumoniae* has been suspended in antiserum, and the presence of the capsule demonstrated by addition of Indian ink. The ink particles are unable to penetrate the capsular material, which therefore appears as a clear halo around the diplococcal cells.

structures are composed of polysaccharides, although some are made of protein or a mixture of both. The capsule of *Bacillus anthracis* is unusual in that it is composed of polyglutamic acid.

The densest external structures are **capsules**. These are easily visualised by negative staining, for example by suspending the bacterial culture in Indian ink. The ink particles cannot penetrate the capsules around the bacterial cells. A capsule thus appears as a clear halo against a dark background. The **quellung reaction** may also be used to demonstrate capsules (Fig. 1.16). *Quellung* is a German word meaning to swell. In this reaction, bacteria are suspended in **antiserum** and the preparation is examined microscopically using oil immersion and diminished illumination. Antibodies in the antiserum may bind to the capsule, giving it a swollen appearance. By using specific antisera raised against particular capsular antigens, bacteria with capsules may be typed using the quellung reaction, since only bacteria with capsules carrying appropriate antigens will appear swollen with specific antisera.

Bacterial capsules are associated with **virulence**. Bacteria that cause **meningitis** are capsulated, and loss of a capsule is correlated with loss of ability to cause disease. When grown in artificial culture, capsulated bacteria give rise to mucoid colonies. Capsulated, virulent strains of *Streptococcus* *pneumoniae* grow as smooth colonies. Strains that have lost their capsules grow as rough colonies. Upon becoming rough, pneumococci also lose the ability to cause disease.

Slime layers are less dense than capsules, and are associated with adhesion. Coagulase-negative staphylococci produce copious extracellular slime, and easily adhere to plastics. Consequently these bacteria, commonly found on human skin, are a frequent cause of infections associated with the use of implanted plastic devices such as artificial heart valves, catheters and valves for the treatment of **hydrocephalus**.

A **glycocalyx** is a fibrous network of polysaccharides extruded by bacteria. Like slime, these are responsible for adhesion. Oral streptococci produce such structures, responsible for the initiation of formation of dental plaque. The most common structures in plaque are **polyglucans** and **polyfructans**, elaborated particularly from sugars present in the human diet. The glycocalyx produced by oral streptococci help them to stick to the enamel surfaces of teeth, and also provide a harbour for other oral microbes, allowing for the accumulation of plaque.

1.7.5   Flagella

Flagella are the bacterial structures that confer motility upon cells. Flagella permit bacteria to escape from their local environment. This allows for the possibility of **chemotaxis**, moving towards a source of nutrients, or away from a chemically hostile environment. However, not all bacteria possess flagella, and as a result, bacteria that lack these structures are generally non-motile. There are, however, exceptions to this rule, including the gliding bacteria, which despite their lack of flagella are capable of motility. Flagella are long structures that may extend several cell lengths, and are composed of a protein helix.

The protein that makes up flagella is called **flagellin**, and is strongly antigenic. It constitutes the bacterial 'H' antigen. 'H' is short for *hauch*, the German word for breath. 'H' antigens are so-called because they cause bacteria to creep over the surface of a solid culture medium in the same way that condensation from breath spreads over the surface of cold glass. Flagellin is associated with a flagellar hook that, in turn, is attached to a basal body. The basal body acts as a rotary motor that causes a gyratory motion in the flagellum. In turn, this propels the bacterial cell forward. At the base of the flagellum are plates associated with the cytoplasmic membrane and involved in rotation. Plates associated with peptidoglycan in the cell wall are responsible for stabilising the flagellar structure. Gram-negative bacteria have an extra plate associated with the outer membrane. In some Gram-negative bacteria the flagella are sheathed, the sheath being continuous with the outer membrane (Fig. 1.17).

Bacteria may have one flagellum or many flagella, and these may be variously arranged. Cells carrying a single flagellum are said to be **monotrichous**; (Greek: *monos*, alone; *trikhos*, hair). Polar flagella are located at one or both ends of the bacterium. Pseudomonads have a single polar flagellum, and are thus described as monotrichous. If flagella are found at both ends, the arrangemet is said to be **amphitrichous** (Greek: *amphi*, at both ends). Flagella that occur in tufts, such as those found in bacteria of the genus *Spirillum*, are known as **lophotrichous** flagella (Greek: *lophos*, a crest). Bacteria that have flagella around their entire surface, such as *Escherichia coli*, are **peritrichous** (Greek: *peri*, prefix for around or about) (Fig. 1.18).

Most bacteria have flagella that extend into the surrounding medium. Spirochaetes are unusual because they have flagella that run along the length of the cell body instead. These structures are called axial fibrils, and they lie just beneath the outer membrane. Axial fibrils are composed of proteins that are genetically very closely related to the flagellins found in other bacteria. They cause spirochaetes to move with a corkscrew motion (Fig. 1.19).

1.7.6   Fimbriae and pili

Many Gram-negative bacteria are covered with fine, hair-like appendages called **fimbriae** (singular: fimbria). The Latin word *fimbratus* means fringed. These structures are so fine that they are only visible using electron microscopy. They play no part in bacterial motility since they may be found in non-motile bacteria. They function as appendages of adhesion, particularly sticking to eukaryotic cells. Consequently they play an important role in helping to establish bacterial infections.

Another appendage of bacterial cells is the **sex pilus** (plural: pili). *Pilus* is the Latin word for hair. The precise function of sex pili remains obscure, but they do act to hold donor and recipient cells together during the act of bacterial conjugation.

Some microbiologists refer to all non-flagellar appendages as pili. They describe those involved in adhesion as type I pili, and sex pili as F pili. This is because the fertility plasmid, F, was found to carry the genes that encode sex pilus formation. Other microbiologists prefer to give these two types of structure separate names to emphasise their different functions.

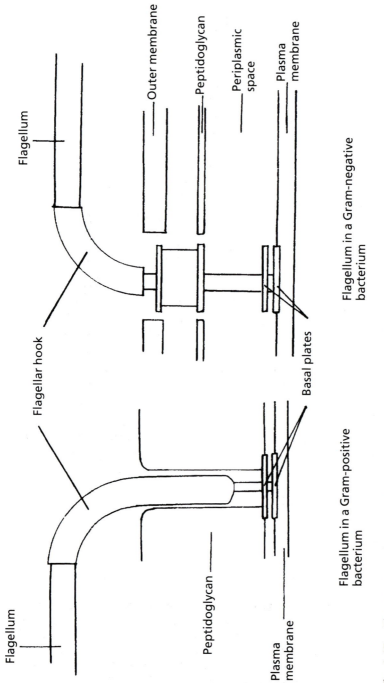

Flagellum

Flagellar hook

Flagellum

Peptidoglycan

Plasma
membrane

Basal plates

Flagellum in a Gram-positive
bacterium

Outer membrane

Peptidoglycan

Periplasmic
space

Plasma
membrane

Flagellum in a Gram-negative
bacterium

Fig. 1.17. The structure of bacterial flagella. One flagellum is from a Gram-positive bacterium (left) and one is from a Gram-negative cell (right).

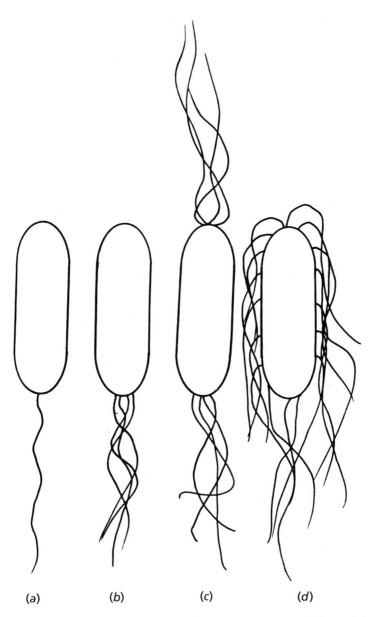

(a)   (b)   (c)   (d)

Fig. 1.18.   Potential arrangements of bacterial flagella. Bacterium (*a*) is monitrichous, carrying a single flagellum. In this case the flagellum is also described as polar, since it is found at one end, or pole, of the cell. Bacterium (*b*) is lophotrichous, carrying a tuft of flagella. The tufts of flagella on bacterium (*c*) are arranged at both ends and are thus described as amphitrichous, and bacterium (*d*) demonstrates peritrichous flagella. In this case the bacterium would appear to be swimming up the page.

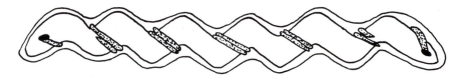

Fig. 1.19. Axial fibrils (shaded) of a spirochaetal bacterium.

1.8 Bacterial spores

The genera *Bacillus* and *Clostridium* have species that are important human **pathogens**. Their **pathogenicity** is considerably enhanced because bacteria of these genera are capable of forming **endospores**. These are often referred to simply as spores. A few other genera of bacteria produce endospores. These include *Sporosarcina*, *Sporolactobacillus*, *Desulfotomaculum* and *Oscillospira* as well as *Thermoactinomyces*, but the spores of the *Bacillus* spp. and of clostridia are the most extensively studied.

Spores are formed in response to adverse environmental conditions. In batch culture they are produced at the end of the logarithmic growth phase, when nutrients become limited and bacterial waste products have accumulated. Spores are highly resistant structures and enable bacteria to survive in more hostile conditions than would otherwise be possible. Spores survive a variety of environmental stresses including high temperatures, desiccation, the chemical activity of many disinfectants and also ultraviolet irradiation. Additionally, spores may survive for many years in conditions of nutrient deprivation. Spores of *Clostridium* spp. have been germinated from samples removed from mummies from ancient Egypt. Bacterial spores have also been germinated from amber thought to be about 4 million years old! Bacterial spores may survive boiling for several hours, and this causes problems when trying to sterilise contaminated solutions and surfaces. Spores are also highly resistant to conventional bacteriological stains and thus require special staining techniques.

1.8.1 Spore structure

The macromolecues in the **spore core** are very similar to those found in vegetative cells of the same species. All the genetic material of the vegetative cell must be located in the spore so that upon germination the genetic complement may be passed on to its progeny. Unlike the vegetative cell, where up to 80% of the wet weight of the cell is water, the spore core is very dehydrated.

This is thought to stabilise the macromolecules present, and may contribute to the resistant properties of spores. DNA inside spores adopts a more compressed conformation, with 11 base-pairs per helical turn compared with 10 base-pairs per turn in vegetative cells. This is considered to contribute to resistance to **ultraviolet irradiation**, although other factors probably also assist in this process.

The endospore core is surrounded by a **plasma membrane** and a **core wall**. The structure of the core wall is similar to that of the vegetative cell wall. Beyond the core wall lies the **spore cortex**. The major cortex component is peptidoglycan, but this has less cross-linking than the peptidoglycan found in vegetative cell walls. Outside the cortex lies the **spore coat**, composed mainly of layers of protein and inorganic phosphates together with small amounts of lipid and polysaccharides. The inner coat is alkali sensitive, the outer coat resistant to alkali. A thin **exosporium** encloses the whole structure in some species. It is composed of protein, carbohydrate and lipid.

Up to 15% of the dry weight of spores comprises **dipicolinic acid** complexed with calcium ions. This is a structure unique to spores. At one time this was thought to be responsible for the heat resistance of spores. The existence of mutants that lack dipicolinic acid but are nevertheless heat resistant disproves this theory. The precise function of dipicolinic acid is not known, but it is thought to stabilise macromolecules (Fig. 1.20).

Spores may be round or oval in shape and may be located in the middle of a bacterial cell (central spores), towards one end (subterminal spores) or at the end of the cell (terminal spores). They may or may not cause swelling of the vegetative cell, known as the sporangium.

1.8.2   Spore formation

Spore formation in bacteria is a highly complex process. It is, nevertheless, relatively rapid, taking about ten hours to complete. Once a spore starts to form, the cell is committed to completing the process. The developing sporangium cannot revert to its vegetative state. The process of spore formation requires considerable coordination of gene expression to be completed successfully. It is therefore the subject of considerable research interest, since it is a topic that lends itself readily to a study of cellular differentiation in prokaryotic organisms.

Microscopically, the first stage in spore formation is the appearance of a faint, clear area in the position that the spore will eventually occupy. This area gradually becomes opaque, and at this point stains intensely. During this period all the DNA in the cell is replicated to provide two complete sets of genetic

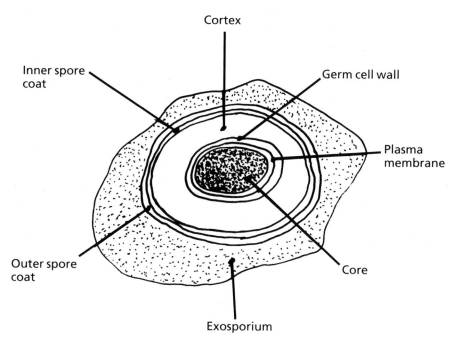

Fig. 1.20.   Structure of a bacterial spore.

information. One set is destined for the spore; the other set remains in the sporangium. The nuclear material is then laid down as an axial filament. Following this, the plasma membrane becomes invaginated, causing the two sets of DNA to become compartmentalised. At this point, a **forespore** develops, and its contents become dehydrated. A double membrane encloses the forespore structure and the peptidoglycan cortex forms between the two membranes. Calcium dipicolinate is synthesised at the same time as the cortex, and is probably incorporated into the core of the spore. Lastly, the spore coat is laid down. Some bacteria also produce an **exosporium**, a loose coat of protein, carbohydrate and lipid. The sporangium containing the mature spore may remain metabolically active for some time, but eventually it lyses, releasing its single spore. This entire process is known as **sporulation** or sporogenesis.

1.8.3   Spore germination

Germination is the process whereby dormant spores are transformed into vegetative cells capable of multiplication but sensitive to environmental stresses. Spores do not germinate inside mother cells, and require a nutrient-rich environment, conducive to the growth of vegetative cells before they will germinate. Even then, some spores require activation by brief exposure to an environmental stress such as heat or low pH. Certain spores, however, may need the presence of only a single chemical, or **germinant**, in the medium to initiate the germination process. Microscopically the germinating spore becomes much less refractile, and easier to stain. It may swell to twice the volume of its dormant structure. Spore wall material, cortex and calcium dipicolinate all become dispersed into the surrounding medium, leaving only the core enclosed by its spore coat. The spore coat then either ruptures, or is absorbed as the germ cell elongates during a period of outgrowth. The **germ cell** then assumes the proportions of a vegetative cell and its nuclear material once again takes up its more relaxed conformation. The germ cell finally divides, yielding two vegetative daughter cells. The process of spore germination is more rapid than sporogenesis, and is usually accomplished in less than 90 minutes (Fig. 1.21).

1.9   Diversity of bacterial nutrition

Bacteria have an extremely wide distribution. They are found from pole to pole. They have been isolated from the icy wastes of the Arctic and the Antarctic, and are found in the waters of hot springs, and near volcanoes on the ocean bed. The diversity of habitats in which bacteria may be found is staggering. It reflects the variety of energy supplies that bacteria can utilise, and also the success with which bacteria have evolved to exploit their environments.

There are bacteria that obtain energy from light; others require a source of chemical energy. Some bacteria utilise inorganic compounds; others require a supply of organic matter. The diversity of nutrition displayed by bacteria is essential for the cycling of elements through biological systems and is a prerequisite for the continued maintenance of life on this planet.

All forms of life require nutrients in order to grow and multiply. Nutrients are the fundamental building blocks from which all living beings are constructed. In order to assemble these building blocks, living organisms require a source of energy. Those bacteria that obtain energy from light are known as **phototrophs**. Literally this means light-nourished (Greek: *photos*, light; *trophe*, nourishment). Bacteria relying upon chemical energy sources are likewise termed **chemotrophs**. The division of bacteria into these two classes is not necessarily absolute, and the energy source exploited may depend upon the

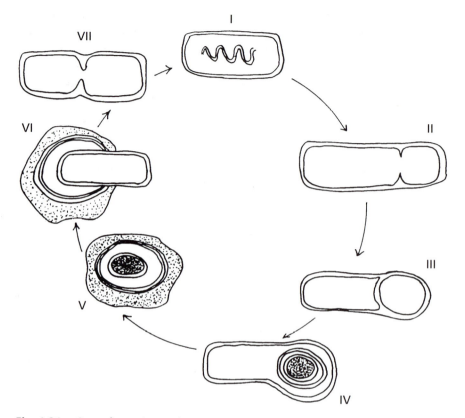

Fig. 1.21.   Spore formation and germination in bacteria. An axial filament (I) then a septum forms (II). The forespore is elaborated (III) and endospore production within the sporangium is completed (IV). The endospore is released from the sporangium (V) and under favourable conditions a vegetative cell germinates from the free endospore (VI). The vegetative cell grows and undergoes binary division (VII).

prevalent environmental conditions. When growing under anaerobic conditions *Rhodospirillum rubrum* is a phototroph and requires a source of light. However, if oxygen is present, then the bacterium can grow in the dark as a chemotroph.

1.9.1   Autotrophic bacteria

All life on this planet is carbon based, and a supply of carbon compounds is essential for the growth and multiplication of organisms. **Autotrophs**, self-nourishing organisms, have the ability to elaborate organic compounds by fixing environmental carbon dioxide. Autotrophic bacteria can be cultured in simple salt solutions, required to provide trace elements for the growing cells. They require no carbon source other than dissolved carbon dioxide. Autotrophs are a vital component of food webs since they produce a renewable supply of organic carbon that is needed by **heterotrophs**, organisms that cannot elaborate all their own organic compounds and that must therefore be nourished by others.

Autotrophic bacteria may utilise light energy, **photoautotrophs**, or may obtain energy from the oxidation of inorganic chemical compounds, **chemoautotrophs**. Photoautotrophs are represented by the purple sulphur bacteria, such as *Chromatium* spp., the green sulphur bacteria such as *Chlorobium* spp. and the cyanobacteria. Chemoautotrophic bacteria include hydrogen bacteria, iron bacteria, nitrifying bacteria and sulphur-oxidising bacteria. Hydrogen bacteria such as *Hydrogenobacter* spp. use hydrogen as an electron donor and oxygen as an electron acceptor to produce water from the reduction of molecular hydrogen and oxygen. Iron bacteria such as *Gallionella* spp. convert ferrous ions to ferric ions. Nitrifying bacteria oxidise ammonia. Those bacteria that oxidise ammonia to nitrite are referred to as nitrosifying bacteria. True nitrifying bacteria can further oxidise nitrite to nitrate. These are exemplified by bacteria like *Nitrobacter* spp. There are bacteria such as *Thiothrix* spp. that are aerobic and obtain energy from the oxidation of hydrogen sulphide.

All of these bacteria obtain energy from the **metabolism** of inorganic compounds. Because of this they are described as **lithotrophs** (Greek: *lithos*, stone); lithotrophs are nourished from stones. Consequently, photoautotrophs may be referred to as **photolithotrophs** and chemoautotrophs are also called **chemolithotrophs**.

In summary, autotrophs have the ability to elaborate all the organic compounds required for their growth and multiplication from carbon dioxide, which they have the ability to fix. Most bacteria are **heterotrophic** and cannot achieve this feat. Instead, heterotrophs rely on an exogenous supply of organic carbon compounds.

1.9.2   Heterotrophic bacteria

Bacteria of the genus *Beggiatoa* were the organisms first described as chemolithotrophs, but many strains are incapable of growth unless they are supplied with organic compounds. Thus, although they obtain their energy from the oxidation of sulphur, and may be considered lithotrophic, they are also heterotrophic because they require a fixed carbon source such as acetate so that they can grow. These bacteria are **mixotrophs** or mixed feeders.

Some photosynthetic bacteria use simple organic compounds such as acetate, formate and methanol as a sole carbon source rather than fixing carbon dioxide. These bacteria are **photoorganotrophs**, and are exemplified by the purple non-sulphur bacteria such as the Rhodospirillaceae when grown anaerobically in the light. Rhodospirillaceae cannot grow anaerobically unless they are illuminated. However, in the presence of oxygen they may grow in the dark as **chemooganotrophs**. Chemoorganotrophs utilise organic compounds as nutrients, and require a chemical energy supply. The vast majority of non-photosynthetic bacteria are chemoorganotrophs.

The purple non-sulphur bacteria display a complex metabolism. Under anaerobic conditions in the light they are photoorganotrophs. Aerobically they behave as chemoorganotrophs. Under low oxygen tension they may display a photosynthetic metabolism simultaneously with an oxidative metabolism. This bestows an enormous metabolic flexibility upon these bacteria. This flexibility allows them to exploit a very wide range of natural environments. A summary of the major groups of microorganisms based upon their nutritional demands is presented in Table 1.2.

1.10   Gene transfer in bacteria

Although bacteria multiply by **binary fission**, an asexual process, they are capable of exchanging genes, and thus maintain a fluidity of hereditary material. Gene exchange in bacteria is not confined within a species, or between closely related species. It readily occurs between bacteria of different genera, and even between bacteria of different families. Occasionally gene transfer can occur between Gram-positive and Gram-negative bacteria. Bacteria have three principal mechanisms for exchanging genetic material: **conjugation**, **transformation** and **transduction**.

Gene transfer has important consequences for humans as well as for bacteria. For example, streptococci are known to harbour a gene that encodes tetracycline resistance. This observation was of no great interest outside the research laboratory until the 1980s. The tetracycline resistance gene spread to a wide variety of bacteria at this time. The spread occurred in bacteria, both Gram-positive and Gram-negative, that are found as part of the commensal flora of the mouth, upper respiratory tract and vagina. The spread of this gene came to prominence when it was found to confer high-level tetracycline resistance upon *Neisseria gonorrhoeae*. It was discovered because people given

Table 1.2 *Nutritional habits of the major bacterial groups*

	The ability to use light as an energy source	The ability to obtain energy from the oxidation of chemicals	The ability to grow using only inorganic chemicals	The requirement for organic chemicals for growth
Autotrophs	Some	Some	All	None
Heterotrophs	Some	Some	None	All
Mixotrophs	Some	Some	Some[a]	Some[a]
Chemotrophs	None	All	Some	Some
Chemoautotrophs	None	All	All	None
Chemolithotrophs	None	All	All	None
Chemoorganotrophs	None	All	None	All
Phototrophs	All	None	Some	Some
Photoautotrophs	All	None	All	None
Photolithotrophs	All	None	All	None
Photoorganotrophs	All	None	None	All

Note: [a] Mixotrophs use either inorganic or organic compounds as growth substrates.

tetracycline to treat cases of **gonorrhoea** failed to respond to therapy. Tetracycline-resistant *Neisseria gonorrhoeae* have rapidly spread around the world, and within five years of being discovered have spread around every continent. They now constitute a world-wide problem in the treatment of gonorrhoea.

1.10.1   Conjugation

Gene transfer in bacteria was discovered during the 1940s. Mutants of *Escherichia coli* were isolated that were unable to grow on a minimal medium without nutrient supplements. These are termed **auxotrophic mutants**. Two auxotrophic mutants with different nutritional requirements were grown in a mixed culture. When plated on minimal medium, progeny were isolated that were able to grow on minimal medium without nutritional supplements. Wild-type genes from one strain complement the mutant genes in the second strain, restoring its ability to grow without nutritional supplement. It was subsequently shown that this process required cell to cell contact, and that genes only moved from a fertile strain (F^+) into an infertile strain (F^-). Some strains, described as **Hfr strains** transfer their chromosomal genes into recipients at very high frequency, hence the name Hfr.

The mechanism of bacterial conjugation became clear with the discovery of F, the fertility plasmid. It encodes the F pilus, and carries all the genes required to replicate itself and pass a copy of itself into a recipient cell with which the donor cell is in intimate contact. The contact is maintained because donor and recipient cells are joined together by the sex pilus. The precise mechanism by which conjugation occurs is still not completely understood. In F^+ strains the F plasmid exists as an independent DNA element. Hfr strains are very efficient at gene transfer because in these strains the F plasmid has become inserted into the chromosome. Integration is mediated by the activity of one of several transposable elements located on the F plasmid.

One F^+ strain may give rise to many Hfr strains, with F inserted at a different location in each Hfr strain. The whole bacterial genome may transfer from an Hfr strain into a suitable recipient if the mating process is not interrupted. It takes about 95 minutes to transfer the entire chromosome of *Escherichia coli* in this manner and the time taken is remarkably constant. The bacterial genome can be mapped by performing **interrupted mating experiments** and using wild-type genes in the Hfr strain to complement mutants in the recipient strain. The wild-type genes transferred from the Hfr donor can compensate for the mutant genes in the recipient, thus allowing **transconjugants** to be isolated on selective media. The mating process is interrupted at different times and mating products are selected. The frequency with which genes appear to transfer at different interruption times gives a measure of how far the mutant gene lies from the point of insertion of F. The map distance is expressed in minutes. It was experiments such as this that illustrated the circular nature of the bacterial genome because different Hfr strains give rise to overlapping map fragments that result in a circular map when assembled together.

The F plasmid may also excise from the chromosome to resume its independent existence. Occasionally the excision process is not accurate, and chromosomal genes may be removed along with the F-derivative plasmid. These are known as F′ plasmids. The F plasmid is not the only plasmid capable of transfer by conjugation. Many large plasmids are self-transmissible into suitable recipients. Small, non-conjugative plasmids may also be mobilised by conjugative plasmids, even though they cannot control their own conjugation.

Conjugation is not a process that is confined to plasmids. Some Gram-positive bacteria harbour conjugative transposons. These are sequences of DNA located in the bacterial chromosome that can transfer to the chromosome of a recipient cell when the donor and recipient bacteria are in intimate contact. This process does not require a plasmid intermediate. The mechanism by which conjugative transposons transfer is, at present, poorly understood (Fig. 1.22).

1.10.2   Transformation

A few genera of bacteria have the ability to take up naked DNA in the process of transformation. The method was originally used to demonstrate that DNA was the hereditary material. DNA isolated from a smooth virulent strain of *Streptococcus pneumoniae* was mixed with cells from a rough non-virulent strain, and the mixture was cultured. Cells growing from the transformation mixture were found to grow as smooth colonies and to have their virulence restored. Natural transformation can occur in both Gram-positive and Gram-negative bacteria including those in the following genera: *Bacillus*, *Streptococcus*, *Acinetobacter*, *Haemophilus*, *Moraxella*, *Neisseria* and *Pseudomonas*. Bacteria that cannot normally take up naked DNA may be induced to do so after treatment with a cold calcium chloride solution to make the cells competent for transformation. The efficiency of artificial transformation is further enhanced by heat-shocking the transformation mixture before culturing the transformants. Precisely how bacteria become transformed is not known, and bacteria may

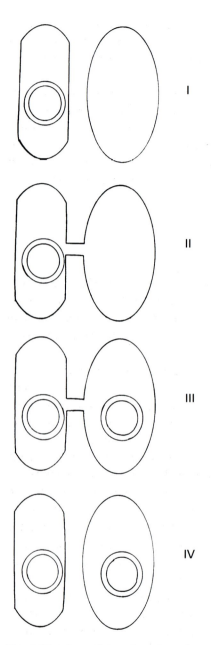

Fig. 1.22. Bacterial conjugation. The donor and recipient cells approach (I) and the donor cell attaches to the recipient via a sex pilus (II). DNA from the donor cell is replicated, and a copy passes to the recipient bacterium (III). Finally donor and recipient cells separate (IV).

have different mechanisms for achieving this type of gene transfer. Circular DNA molecules such as plasmid DNA transform with much greater efficiency than does linear DNA. Small plasmids are generally easier to transform than large plasmids.

Transformation is widely exploited in genetic engineering to introduce foreign DNA into bacteria. The DNA fragment to be studied or exploited is typically inserted into a specialised small plasmid, known as a **cloning vector**, using an enzyme called DNA ligase. The **recombinant DNA** is then used to transform competent bacterial cells, and the transformants are cultured on an appropriate selective or indicator medium prior to further studies or the exploitation of the novel recombinant product.

1.10.3   Transduction

Bacterial genes may be transferred from strain to strain using bacteriophages as **vectors**. This process is referred to as transduction. There are two fundamentally different transduction processes, reflecting the different ways in which bacteriophages behave: **generalised** and **specialised transduction**.

Generalised transduction   The most usual outcome of bacteriophage infection of a cell is that the bacteriophage replicates within the cell, causing disruption of the cellular components and ultimately cell lysis, allowing release of new bacteriophage particles. These particles may subsequently infect other susceptible cells, completing the **lytic cycle**. Occasionally during the lytic process, the bacterial genome becomes fragmented and some bacterial DNA segments may become packaged into **nascent bacteriophage particles** as well as bacteriophage DNA. Bacteriophages that contain bacterial DNA do not have room for all the bacteriophage genome, and so they are defective. They are, however, capable of infecting other bacterial cells, introducing the bacterial DNA that they carry into a new host. The newly introduced DNA may then recombine with the genome of its new host, thereby becoming stably incorporated into the cell. This process is generalised transduction.

Only a very small proportion of the bacterial genome may become incorporated into the defective bacteriophage head, usually less than 1% of the total DNA. Most particles made during the lytic cycle are normal, non-transducing bacteriophages. Transducing bacteriophages represent a minority population. These considerations, together with the random incorporation of bacterial DNA into transducing bacteriophages, makes the probability of transferring a particular gene very low (Fig. 1.23).

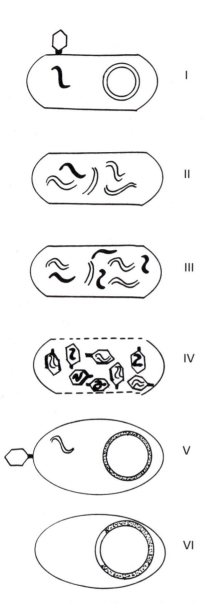

Fig. 1.23.   Generalised transduction. A bacteriophage infects a bacterium. (I). The bacterial DNA is broken down into fragments (II). The bacteriophage genome is then replicated (III). Bacteriophage particles are formed, and packed with nucleic acid. In some instances, bacterial DNA will become packaged into a bacteriophage head to produce a transducing bacteriophage. These are released from the bacterium by lysis of the host cell (IV). A new strain is then infected with the transducing bacteriophage (V) and DNA from the original host is inserted into the genome of its new host (VI).

Specialised transduction   **Temperate bacteriophages**, as well as caus-ing lytic infection, are capable of forming more stable relationships with their bacterial hosts. Upon infection, instead of causing immediate replication and lysis, the DNA of temperate bacteriophages may become incorporated into the bacterial genome as a **prophage**. Prophages may encode other products in addition to those responsible for bacteriophage functions. For example, diph-theria toxin is encoded by a prophage, and only those strains of *Corynebacterium diphtheriae* that carry the toxin-encoding prophage can cause **diphtheria**. Bacteria harbouring prophages are known as **lysogens**.

Prophages may exist in stable harmony with their hosts for long periods, but upon exposure to particular environmental stresses such as elevated tem-perature, ultraviolet irradiation and particular chemicals, a lytic growth cycle is induced. When the bacteriophage DNA is excised from the host genome, its excision is not always precise, and some bacterial DNA may be removed at the same time as the prophage DNA. Bacterial genes may thus become packaged into defective bacteriophage particles, which may then infect suitable recipi-ents. Upon injection of such bacterial genes into a new host, they may become stably incorporated either by the initiation of a subsequent lysogenic infec-tion, or by recombination with the genome of the new host cell. This process is called specialised transduction because each prophage has a particular site of insertion in the bacterial genome. In consequence, only bacterial genes adja-cent to the insertion site of the prophage can be transferred by this process. These genes are transduced at higher frequency than are the genes whose transfer is accomplished in generalised transduction (Fig. 1.24).

1.11   The nature of viruses

Viruses are very simple structures. A virus is essentially composed of a central core of nucleic acid surrounded by a protein coat. The nucleic acid present in a virus is either DNA or RNA, but viruses never contain both types of nucleic acid. Some viruses are enclosed within a lipoprotein or glycoprotein mem-brane called an **envelope**. The core, protein coat and, if present, the envelope are referred to as a **virion**. The simplicity of the basic virus structure, however, is not reflected in the way in which they replicate, and they require other cells to perform this function. It may be asked whether viruses are alive or not, and there is no satisfactory answer to this problem. One of the difficulties raised by this question is how to define what constitutes life. If the definition of life is limited to the ability to replicate, then viruses may be considered to be alive when present in a host cell. However, outside their host cells, viruses are

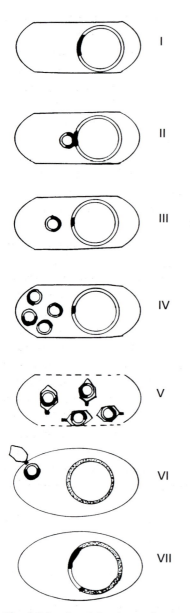

Fig. 1.24. Specialised transduction. A prophage exits as a stable entity in a lysogenic bacterium (I). Occasionally, the prophage sequence is excised from the host genome (II). This process is not always precise, and some bacterial DNA may also become excised (III). This excised nucleic acid is replicated (IV), and is packaged into new bacteriophage particles that are released by lysis of the host bacterium (V). A transducing bacteriophage infects a new bacterium (VI), where the DNA, including sequences from the original host, may become stably incorporated into the genome of the new host, initiating a subsequent lysogenic infection (VII).

merely collections of macromolecules that are in a more or less inert state. When apart from their host cells, viruses can hardly be described as alive.

Without attaching to and penetrating a suitable living host cell, viruses cannot replicate. Thus, in order to survive and evolve, viruses require a population of susceptible hosts. Outside their hosts, viruses do not generally endure for very long. One of the principal reasons for this is that the protein and nucleic acids of viruses may be easily denatured, rendering the virus particle ineffective. This property is a fundamental weakness of viruses, and it has been exploited to rid the world of **smallpox**, a devastating infection caused by the variola virus. Edward Jenner's work in the eighteenth century started the process of **vaccination** when he inoculated a boy named Phipps with material from a **pustule** on the hand of a milkmaid who had **cowpox**. Later, he inoculated the same boy with smallpox, but this failed to cause disease. As a result of this success, a group of children were vaccinated with material obtained from cows infected with cowpox, and when these children were later challenged with smallpox they too were protected from the disease. These experiments formed the basis of modern vaccination programmes. The very name vaccination is derived from *vacca*, the Latin for a cow. These days, smallpox vaccination uses vaccinia virus. The exact nature of this virus is still debated, but it appears to be a hybrid between the cowpox virus and the smallpox virus. Mass vaccination programmes strive to produce large numbers of individuals in populations who are protected from infection. With virus infections, this limits the number of people in whom the virus can replicate. In turn, this will limit the spread of the virus. The World Health Organization (WHO) initiated a vaccination programme against smallpox in 1967. This programme was so successful that by 1977 the last natural cause of this horrendous disease was reported. Because of the vast numbers of individuals who were vaccinated in countries where smallpox was once a major killer disease, there were so few susceptible individuals that the virus had nowhere to replicate. In turn, this caused the virus to become extinct in the wild. Boosted by the success of this programme, WHO now have programmes of eradication for other virus infections, such as **poliomyelitis**. The partial success of such programmes was illustrated recently. In 1994, it was announced that poliomyelitis had been eradicated from the Americas.

Viruses have to rely totally on other living cells for their replication and thus they are described as **obligate intracellular parasites**. One infectious particle enters a cell, and replication of the virus will give rise to hundreds, or in some cases, thousands of new virus particles, sometimes described as the progeny. In most cases, a virus replication cycle results in the death of the host cell. It is the death of host cells that give rise to disease. For example, hepatitis viruses kill the cells of the liver, influenza virus kills the cells lining the respiratory tract, and human immunodeficiency virus, or HIV, kills cells of the immune system. It is a grim irony that HIV kills the very cells that protect us from other infections. Most species of animals, plants, fungi and protozoa are susceptible to infection with viruses. Bacteria are also susceptible to virus infection, and the viruses that infect bacteria are termed **bacteriophages**. Most of the following descriptions of viruses are concerned either with bacteriophages or with animal viruses, but the effect of plant viruses on human society should not be forgotten.

In addition to being obligate intracellular parasites, viruses also exhibit **cell tropism**. This means that there is only a limited range of cell types that will support the growth of a particular virus. The phenomenon of tropism is usually determined by the presence or absence of specific **anti-receptors** on the surface of the virus and **receptor** sites on its host cells. Cell tropism not only determines the cell type to which a virus will attach, but also which species of animal, plant, fungus or bacterium will be susceptible to that virus. It is for this reason that dogs do not suffer from poliomyelitis, and that humans do not suffer from canine distemper. Poliomyelitis virus will naturally attach to and infect only cells of primate origin, because other types of cell lack the appropriate receptor sites. However, if poliomyelitis viruses are carefully injected into mouse cells, the virus will replicate to produce new progeny virions. This shows that the replication of poliomyelitis virus is not limited to primate cells, but its ability to infect cells is thus restricted. It is usually a protein or, with enveloped viruses, a lipoprotein or glycoprotein molecule on the surface of the virion that attaches to the host cell's receptor site to initiate the virus infection of the cell. For instance, the haemagglutinin molecule on the surface of the influenza virus attaches to sialic acid residues on the surface of respiratory endothelial cells. Similarly, a glycoprotein molecule, gp 120, on the surface of the human immunodeficiency virus attaches to the CD4 receptor of specific cells of the human immune system.

In some instances, virus replication does not lead to the death of a cell, but to proliferation of its host. This is when a normal cell becomes **transformed** into a **tumour cell** as a result of virus infection. Proteins that are encoded by viruses alter the normal biochemistry of the cell, and this can lead to transformation of normal cells into tumour cells. The commonest cause of death in cats is road accidents. The second most frequent cause of death is leukaemia, and this is caused by the feline leukaemia virus, similar in structure to the human immunodeficiency virus. Both viruses belong to a group known as the retroviruses. These viruses have the ability to make a DNA copy of their RNA genome and to insert this DNA copy into the genome of the host cell in the process of **integration**. To facilitate this activity retroviruses carry within the virion an enzyme called **reverse transcriptase**. Upon infection of a new host this enzyme makes a DNA copy of the virus RNA.

The replication of a virus within a cell is not susceptible to the effects of antibiotics, and nor is the virus-induced transformation of normal cells into tumour cells. One of the principal reasons for this is that viruses rely on many of the functions of the host cell to effect their continued survival, and so most chemicals that inhibit viruses are also very toxic to normal cells. Therefore there are very few agents that can be used to treat virus infections. Consequently, viruses include some of the most important life-threatening infectious agents in developed countries as well as in the Third World.

1.11.1   The development of virology as a science

Historically, virus infections have been recognised since the dawn of civilisation. The ancient Egyptians were stricken with smallpox, caused by variola virus, and suffered **cold sores**, as a consequence of herpes simplex virus infection. There are even pictures from the ancient Egyptian civilisation that depict people with withered limbs, such as is typical of poliomyelitis. Despite this, the science of virology is relatively modern, even with respect to other aspects of microbiology. During the development of microbiology, attempts were made to harvest infectious particles by filtration. Filters were designed with pore sizes small enough to trap fungi, protozoa and even bacteria. However, there exist infectious particles that are capable of passing through filters, and these were referred to as **filterable viruses**. The word *virus* in Latin means a slimy liquid, venom or poison. The prefix 'filterable' was soon dropped, and infectious agents smaller than bacteria were simply called viruses. Viruses are so small that they cannot be seen in the light microscope. During the early part of the twentieth century, viruses were regarded merely as minute versions of other microorganisms, incapable of independent replication. The advent of animal tissue cultures and the development of the electron microscope were to help change this view. Improved chemical and analytical techniques showed viruses to be distinct entities, quite unlike other microorganisms.

During the 1960s and 1970s the new science of virology underwent a rapid expansion as new techniques became available to biologists. During the 1960s, biochemists used bacteriophage systems to elucidate the way that viruses synthesise proteins and replicate their nucleic acids, and in the 1970s and 1980s molecular biologists determined the **nucleotide sequence** of many virus genomes. The genomes of simple, single-stranded RNA bacteriophages such as f2 and Qβ were among the first genomes to have their nucleotide sequences determined, and the single-stranded DNA bacteriophage, M13, has been widely used in cloning experiments to generate DNA templates for the determination of many gene sequences. Such studies have made a huge impact on our understanding of molecular biology. More recent studies in virology have included the discovery of new viruses, including HIV, responsible for AIDS, and human herpesviruses 6 and 7 that have a complex biology but do not cause severe clinical infections.

Viruses are a major cause of disease in humans. They are responsible for more General Practitioner consultations than any other condition. Also, as intriguing biochemical tools, they continue to be widely used by molecular biologists in academic research. Recently, beneficial effects of viruses have started to be explored. There are studies that indicate that viruses may have a novel and innovative role in human **gene therapy**. Viruses are being proposed as vectors to deliver normal human genes into cells that carry defective copies of particular genes in the hope that the healthy gene will carry out its normal function, thus alleviating the deficiency caused by the abnormal gene. **Cystic fibrosis** is a disease where considerable progress has been made in this field.

1.12   Structure and composition of viruses

Unlike eukaryotic or prokaryotic cells, viruses contain only one type of nucleic acid. The nucleic acid of a virus may, however, be linear or circular, single- or double-stranded, and may be either DNA or RNA. The size of the virus genome ranges from a few thousand nucleotides up to 250 000 nucleotides. The nucleic acid carries virus-specific genes that code for non-structural proteins, essential for the functioning of the virus. In addition, the virus genome encodes structural proteins that constitute part of the virion. **Structural proteins** are the building blocks of the virus, and **non-structural proteins** are those that are found in infected cells. The enzymes that are necessary for virus replication are examples of the non-structural proteins encoded by the virus genome. However, many viruses carry enzymes as structural components of the virion. For example, the human immunodeficiency virus carries the reverse transcriptase enzyme that is necessary for generating a DNA copy of the virus RNA as part of the virus particle. Such proteins constitute a third class of virus protein, responsible for an enzymatic function within an infected cell, but also part of the structure of the virus particle.

The virus nucleic acid is in close contact with, and is surrounded by, a protein coat, referred to as the **capsid** or shell. The capsid is composed of protein subunits called **capsomeres**. In most viruses, it is the capsomeres that constitute the major protein component of the virion. The capsid offers excellent protection to the nucleic acid that it encloses. Electron microscopic and X-ray diffraction studies have been used to study the structure of viruses. These show that virus capsids can be assembled according to different patterns. Viruses with an **icosahedral** symmetry, such as adenoviruses, are regular bodies with 20 faces of equal size. Helical viruses, such as the tobacco mosaic viruses, have capsomeres aligned in a spiral pattern. Some viruses, such as the bullet-shaped rabies virus, have a combined structure. Others, like the T-even bacteriophages, have a complex structure. In many viruses, the capsid is surrounded by an envelope. Virus envelopes may be tight-fitting or baggy, and they are composed of virus-encoded glycoproteins embedded in a membrane. The envelope is derived from either the nuclear membrane or the plasma membrane of the host cell by a process of budding.

Icosahedral viruses such as adenoviruses, turnip yellow mosaic virus and herpesviruses have a regular polyhedral shape, with 20 triangular faces, 30 edges and 12 corners. The number of capsomeres present is constant for a particular virus, but varies between virus types. For instance, the adenovirus capsid has 252 capsomeres and herpesviruses have 162 capsomeres whereas the turnip yellow mosaic virus has only 32 capsomeres. Adenoviruses also carry characteristic fibres that protrude from the icosahedral capsid.

Helical capsid viruses such as tobacco mosaic virus and influenza virus resemble long rods that can be either flexible or rigid depending upon the type of virus. The capsid is a hollow cylinder that has a helical structure which surrounds the nucleic acid. Helical capsids usually have a very close and intimate association with the virus nucleic acid. Indeed, such structures are often called **nucleocapsids**. The condensed nucleic acid within an icosahedral capsid has less contact with the proteins of the shell than do the nucleocapsids of helical viruses (Fig. 1.25).

The 'lunar landing module' structure of the T4 bacteriophage is an example of a complex symmetrical architecture that is unique to the T-even bacteriophages. These bacteriophages have an icosahedral head, containing the virus DNA, attached to a helical sheath or tail that may or may not be contractile. At the base of the tail is a plate. In addition, some T-even bacteriophages have tail fibres attached to the base plate. These fibres play an important role in the infectious cycle of the T-even bacteriophages.

For some viruses, such as tobacco mosaic virus and enteroviruses such as poliomyelitis virus, the capsid structures described above are unmodified,

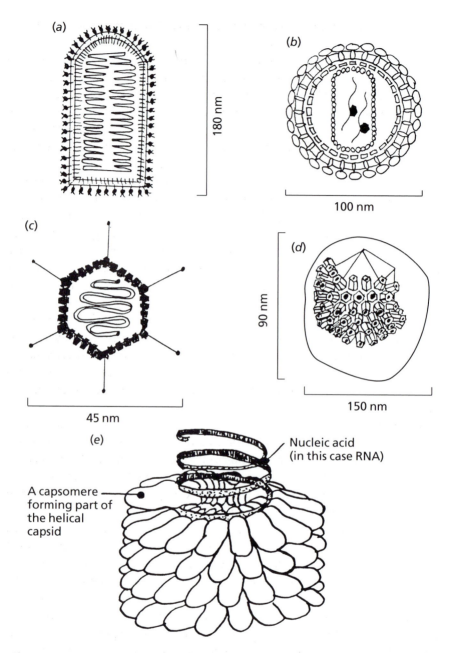

Fig. 1.25. Structures of various viruses. (*a*) A section through the rabies virus. (*b*) A section through the human immunodeficiency virus. (*c*) A section through an adenovirus. (*d*) A cut-away figure of herpes simplex virus and (*e*) a cut-away of the plant tobacco mosaic virus.

and they represent the whole virion structure. However, for many viruses, including the rabies virus, HIV, and viruses of the herpesvirus family, the capsid is in close or loose association with a virus envelope. This is composed of virus-encoded glycoproteins and host cell-derived lipids. The envelope plays an important role in virus attachment and entry into a new host cell, and with the non-lytic budding of virus particles from the host cell. Virus envelopes contain many glycoproteins, not all of which are essential for virus replication and formation. Others are essential since they provide a receptor/anti-receptor contact that is necessary for initiating a virus infection.

Since either the capsid proteins in viruses such as poliomyelitis virus and tobacco mosaic virus, or the envelope proteins in viruses such as influenza, HIV and the herpesvirus family have an important role in initiating virus infections, they have been the subject of intense investigation by those immunologists and molecular biologists who are striving to produce improved virus vaccines. The immunity that follows infections such as **influenza** results from the induction of antibody production in the immune individual.

Antibodies raised against the influenza virus combine with the haemagglutinin molecules on the virion to neutralise them, rendering the virus non-infectious. Inoculation of individuals with purified haemagglutinin molecules, or fragments generated from such molecules causes the production of antibodies against the haemagglutinin, and these afford the immunised person protection from infection with influenza virus. Because the vaccines use only a part of the infectious agent, rather than the whole structure, they are referred to as **subunit vaccines**. The successes of influenza vaccination programmes have stimulated attempts to mimic this procedure with other virus structural proteins. For example, the G-protein spike in the envelope of the rabies virus and the surface glycoprotein, gp 120, found in the envelope of HIV are being investigated.

Whilst many attempts to produce subunit vaccines are at an experimental stage, there is one subunit vaccine that is regularly used to **immunise** members of the dental, medical and paramedical professions as well as other individuals at risk of contracting hepatitis B. It exploits recombinant DNA technology to produce the S-protein of the hepatitis B virus to provide the antigenic stimulus. This vaccine has the advantage that it does not require the person being vaccinated to be exposed to the whole hepatitis B virus. This and other viruses may be found in blood, and can often be collected from samples pooled from a number of individuals. With such procedures, used in early vaccine production for example, there is always a risk that other infectious agents will be purified along with the desired virus, and that these agents may not be inactivated. It was in this manner that many haemophiliacs were infected with HIV as a result of contamination of batches of the blood clotting factor, **factor VIII**, that they require. In consequence, many haemophiliacs have died of AIDS.

2

Handling microbes

2.1 Safe handling of microbes

The vast majority of microbial life on Earth is harmless to humans, and many microorganisms have beneficial effects. Since the dawn of civilisation, humans have harnessed microbial fermentations to make bread and alcoholic beverages, and to prolong the life of food. Today, technologists are exploiting microorganisms in the pharmaceutical industry, for food production, mineral extraction, the oil industry and in agriculture. There is hardly any aspect of modern life that is not touched by microbiology. However, a small minority of microbes do cause disease, and a minority of disease-causing microbes can cause fatal infections. Some, such as the human immunodeficiency virus, the cause of AIDS, may take several years to exert their lethal effect. Others such as *Neisseria meningitidis*, the cause of meningococcal meningitis, can kill within hours of the first symptoms of the disease.

When working with microbes, care must be taken to ensure that laboratory cultures do not escape to cause laboratory-acquired infections or to pollute the environment. Equally, it is important to ensure that laboratory cultures do not become contaminated with unwanted extraneous organisms from the environment. If care is not taken to avoid contamination of laboratory cultures, then the results of microbiological experiments are not reliably reproducible, and the data obtained would be unreliable. It is impossible to tell whether the observations made in such experiments are due to the properties of the desired organism, or arise from the activity of a contaminant. Working with microorganisms demands a professional approach to safety, and also a degree of technical competence. Research workers and the personnel in diagnostic microbiology laboratories may require a programme of vaccination to provide additional protection against microbial infection potentially acquired because of the nature of their work.

Prior to the start of experimental procedures, all apparatus and media to be used must be sterilised. All unwanted cultures and potentially contaminated equipment must be appropriately dealt with after laboratory work has been completed. Cultures should be sterilised before being disposed of. Live cultures should never be poured down the sink. Equipment should be either sterilised or decontaminated as appropriate. At the end of the working day, all work surfaces should be wiped clean with a disinfectant solution.

During an experiment, it is important to observe *good microbiological practice*. Clothing should be protected at all times by a clean laboratory coat, which must be worn properly fastened. An open laboratory coat protects only the wearer's back. Laboratory coats should preferably be white so that contamination can be easily spotted. Laboratory coats should not be allowed to leave the laboratory without being sterilised. They should even be sterilised before laundering. Many microbiological experiments involve the use of potentially dangerous organisms. These only become dangerous when handled incorrectly or carelessly. However, *all* cultures should be regarded as potentially dangerous.

Cultures should be manipulated using a good *aseptic technique*. This comprises practices designed to minimise the risk of contamination. Liquid cultures should be handled with the minimum of shaking to reduce the risk of **aerosols** forming and dispersing. Cultures should only be exposed to the air for as long as is necessary to make observations or perform manipulations. Agar cultures in Petri dishes should always be kept upside-down so that any condensation forming on the lid does not fall onto the surface of the culture. Mouth pipetting is *NEVER* permitted in microbiology laboratories, even for dispensing sterile distilled water. Bottle tops, test-tube caps and flask stoppers such as cotton wool bungs must never be placed on the work surface, where they may easily become contaminated. With a little practice, even large bottle tops can be easily manipulated in the crook of the little finger, leaving the rest of the hand free for manipulation of cultures, loops, and so on (Fig. 2.1). Open cultures should be manipulated close to the flame of a Bunsen burner, and they should be held so that they face away from any convection currents that may carry aerial organisms. The necks of culture vessels and reagent bottles should be passed through a hot Bunsen burner flame upon removal of the stopper, bung or bottle top, and again after the manipulation is complete and before the cap or stopper is replaced. This will kill any contaminating organisms on the neck of the vessel before the manipulation, or any of the culture contaminating the mouth of the vessel after manipulation and before

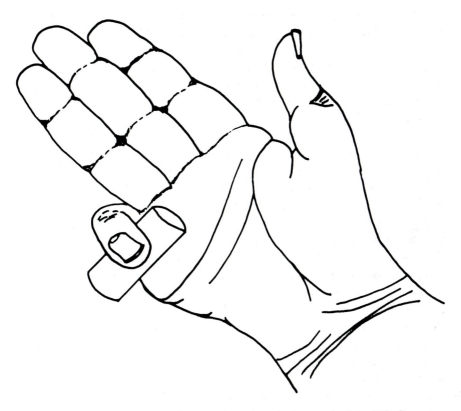

Fig. 2.1.   Handling the stopper from a tube culture in the crook of the little finger. This leaves the other fingers and thumb free to manipulate other objects.

replacing the cap or stopper. When cotton wool is used as a bung, care must be taken because the material is flammable.

Fire is a major hazard in microbiology laboratories. Lighted Bunsen burners must never be left unattended, and long hair must be restrained. This avoids both the fire risk and the risk of accidental contamination by or of cultures. Bacteriological staining often employs flammable solvents such as acetone or alcohol. Great care must be employed when handling these, particularly near Bunsen burners.

A very widely used tool in microbiology is the metal bacteriological loop. Before this is employed it is heated to red heat in a Bunsen burner flame to sterilise it. It is then allowed to cool before use to avoid damaging the culture, or spattering live microbes into the atmosphere. It is essential that the loop is flame sterilised after use as well, in order to kill any culture cells that adhere to it. When heating a contaminated loop, care must be taken to avoid dispersing

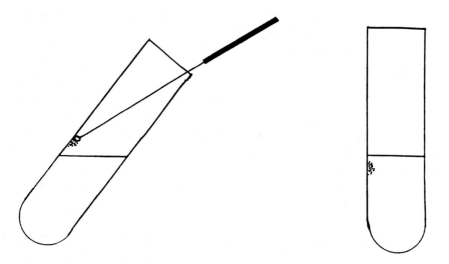

Fig. 2.2.   Inoculation of a liquid culture. By tilting the tube and inoculating the glass at a point that would normally be below the fluid level, the risk of producing aerosols is minimised.

live organisms. To do this, the loop is first introduced into the cooler part of the flame, and only when it has warmed up is it placed in the hottest part of the flame.

When a liquid culture is to be inoculated, a charged loop should not be placed directly into the broth. Rather, the culture should be tilted slightly and the **inoculum** rubbed on the side of the vessel. The inoculum should be emulsified in the liquid that clings to the side of the tube in a position where it will be submerged when the culture is placed in its upright position for incubation (Fig. 2.2).

If accidental spills of cultures do occur, any liquid should be quickly and carefully prevented from spreading by the application of absorbent paper towels or cotton wool. This contaminated material should then be placed in a suitable receptacle for later sterilisation. The affected work surface should be thoroughly cleaned using a suitable disinfectant, active against the microbe being manipulated. When spills occur, it is important that thorough hand-washing is practised immediately after the spilled material is cleared up and before work is resumed.

The mouth is an important route of infection. Eating, drinking and smoking are forbidden in microbiology laboratories. Fingers, pens, etc. should never be placed in the mouth, and lipsticks should not be applied in laboratories, to minimise the risk of orally acquired infection or chemical poisoning.

The application of other cosmetics in microbiology laboratories is also forbidden to reduce the risk of infection through the skin.

The skin provides a major anatomical barrier to infection. Any cuts and grazes breach this defence, and so should be covered before microbiological manipulations are started. The use of rubber gloves helps to protect against accidental infection. If cuts or grazes occur during experimental work, medical advice must be obtained. An open cut or graze provides an unusual portal of entry for infection, and a resulting infection may well not show typical clinical symptoms because the infectious agent has been allowed access to the body through an unusual route. At the end of laboratory work and after laboratory coats have been removed it is important to wash hands thoroughly to remove contaminating microorganisms, preferably using sinks equipped with knee-controlled taps that do not require manual operation. These reduce the risk of contamination and cross-infection. Outdoor clothes and bags should not be brought into the laboratory, and notebooks, etc. should be kept to a minimum. Live cultures should never be placed on or manipulated over notebooks that will be taken out of the laboratory.

There is a tiny minority of microorganisms and viruses that are highly infectious, and whose infections are potentially fatal. Cultures of these microbes require much greater degrees of containment than used in most microbiology laboratories. Handling such agents requires the use of safety cabinets and more protective clothing that may cover the entire body. It may even require special isolation facilities in purpose built containment laboratories.

2.2   Sterilisation and disinfection

One of the most important aspects of the control of microorganisms is the ability to inhibit or kill unwanted microbes. This may be achieved in several different ways, and different degrees of inhibition can be achieved. If a compound or treatment results in the death of microbes, it is said to be **bactericidal** or **fungicidal**. Agents that destroy viruses may be described as **viricidal**, but this pre-supposes that viruses are independent life-forms. In the popular literature, these agents are frequently referred to as **germicides**, perhaps because they kill all known germs. In contrast, if a compound merely prevents microbes from growing it is said to be a **static** agent. Many of the chemicals used as agents of sterilisation or disinfection are **cidal**, killing the microbes against which they act. However, food preservatives must not be toxic to humans, and so chemicals used as food preservatives are frequently static in their antimicrobial activity. Mercury ions are also **bacteriostatic** rather than bactericidal, even at relatively high concentration. These terms are not applied simply to disinfectants, and may also be used to describe **antibiotics**. These are special antimicrobial agents that are produced by microorganisms to inhibit or kill other microorganisms. Chloramphenicol is an antibiotic that inhibits bacterial cells by interfering with protein synthesis. Upon removal of the drug, however, the bacterial culture resumes normal growth. Thus chloramphenicol is a bacteriostatic agent. In contrast, streptomycin is a bacterial protein synthesis inhibitor that is bactericidal since bacteria cannot recover from the effects of exposure to this agent. The use of antibiotics and antimicrobial chemotherapeutic agents is a special example of the control of microorganisms but, the topic is beyond the scope of this book (Fig. 2.3).

Sterilisation and disinfection are both processes used to limit microbial activity. There is often some confusion about the use of these terms. **Sterilisation** involves the killing of all living cells, including spores, as well as viruses and **viroids** from an object, surface or habitat. It also involves the destruction of poorly characterised infectious agents such as those that cause **spongiform encephalopathies** such as **scrapie** in sheep, bovine spongiform encephalopathy (BSE) in cattle and Creutzfeldt–Jakob disease in humans. Sterilisation is an absolute phenomenon. An object may be either sterile or non-sterile. **Disinfection** is a more nebulous term. It may be considered to be the removal or inhibition of microorganisms that are likely to cause disease. Some groups of people are at a much greater risk of developing infectious diseases than others. Vulnerable groups include newborn babies, pregnant women and the immunocompromised. Thus, defining disinfection also involves defining the group of people at risk. **Antisepsis** is the removal of potentially infective organisms from living tissue. Because antisepsis should avoid damaging host tissue, **antiseptics** are generally more mild in action than are **disinfectants**, used for the disinfection of objects or habitats. Neither antisepsis nor disinfection results in the killing or removal of all life-forms as does sterilisation. **Sanitisation** is another term used in this context, and represents the reduction in the microbial load of an object or surface to levels that are considered safe for public health. This is a very imprecise term and is affected by many factors. A wide variety of physical and chemical agents are employed in sterilisation, disinfection, antisepsis and sanitisation.

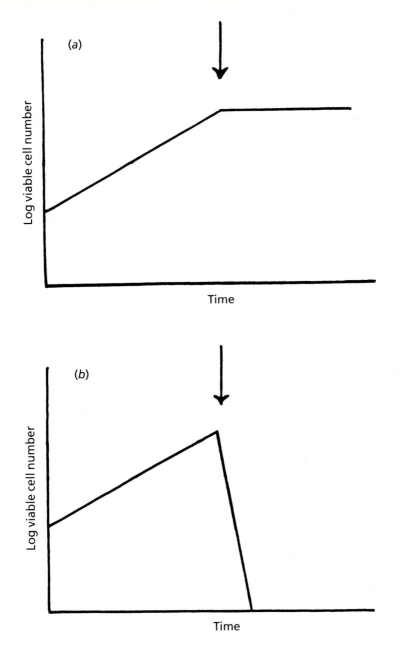

Fig. 2.3. The effects on growth of (a) a bacteriostatic and (b) a bactericidal agent. The viable count in (a) remains constant after the addition (at arrow) of chloramphenicol, and would rise again if the cells were resuspended in a chloramphenicol-free medium. (b) After the addition (at arrow) of streptomycin, bacterial numbers decline rapidly. Other bactericidal agents may kill bacteria more slowly than streptomycin does; it kills a culture in less than 30 seconds. If the cidal agent acts more slowly, then the slope of the line will be less steep following addition of the drug.

2.2.1 Physical agents used for sterilisation and disinfection

Since time immemorial humans have used fire as a purging agent. In biblical times, articles that had been contaminated by lepers were consigned to the flames. This, we now know, is an effective means of sterilising such material. The flaming of a bacteriological loop has the same effect. It is common practice in microbiology laboratories to remove contaminating organisms from the surface of instruments by dipping the instrument in alcohol and burning off the spirit. This, however, kills only the heat-sensitive vegetative cells on the object, since the temperatures achieved in this process are insufficient to kill any spores that may be present.

There are different ways of defining the effectiveness of heat for killing microbes. The **thermal death point** defines the minimum temperature at which cells are killed following an exposure of ten minutes' duration under specified conditions. This is not a precise value, since it depends upon the number of organisms in the culture at the time of exposure. However, it is easily and quickly determined, and it provides an indication of the heat resistance of microbes. It should be remembered that the vegetative cells of sporing bacteria have a much lower thermal death point than do the heat-resistant spores they produce.

An alternative measure of the effect of temperature on the viability of microbes is the **thermal death time**. This is defined as the time taken at a given temperature to kill all the microbes in a suspension held under defined conditions. As with the thermal death point, this measure is imprecise and depends upon the number of microbes in the culture at the time of measurement.

To overcome the imprecision of the thermal death point and the thermal death time, the **decimal reduction time** or **D value** has gained wide acceptance. The D value is the time required to kill 90% of microorganisms at the stated temperature and in the specified matrix. The D value is usually written with a subscript denoting the temperature to which it applies. When the number of viable microbes is plotted against time on semi-logarithmic graph paper as shown in Figure 2.4, the D value is seen as the time taken for the population of cells to drop by 90%. Graphically this is seen as a reduction through one log cycle. This represents a tenfold reduction in the number of viable cells present, hence *decimal* reduction time.

In commercial food canning operations, great care is taken to ensure that the product is free from *Clostridium botulinum*, the causative agent of **botulism**.

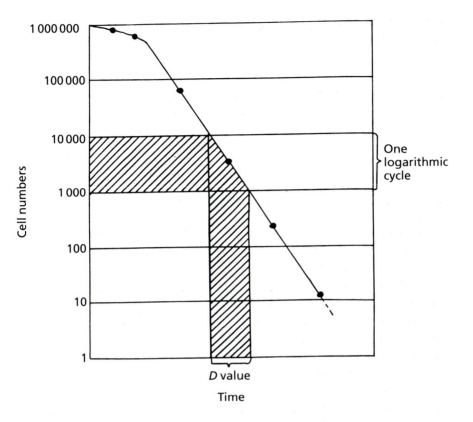

Fig. 2.4. Determination of a D value using a graphic method. The number of viable microbes surviving at the desired temperature at different time points are plotted on a semi-logarithmic graph. The data used for this figure are given in the following table:

Time at 115 °C (minutes)	Number of survivors
0	1×10^6
1	8.9×10^5
2	6.2×10^5
4	7.1×10^4
6	3.6×10^3
8	2.5×10^2
10	12
12	<1

The D value is then the time taken for the viable count to drop through one logarithmic cycle in a portion of the curve where the loss of viability is exponential, and in this case the D_{115} value is equivalent to approximately 1.6 minutes.

Canned food is held at the required temperature for a time equivalent to 12 × D value of *Clostridium botulinum* spores in that particular food. This is to ensure that the viable spores within the food are reduced by a factor of 10^{12}. The chance of even a single spore surviving for this period is so small as to be negligible, and so the food is considered safe to eat. For just one spore to survive such a process, the food prior to processing would have had to contain 10^{12} spores.

Microbes are killed by exposure to high temperature. The combination of heat and water is more effective than dry heat. Moist heat kills by the combined effects of oxidation and coagulation of cellular components. Dry heat relies much more upon oxidation than upon coagulation to cause its lethal effects. To illustrate the difference between moist and dry heat: it takes only five minutes to kill the spores present in a typical culture of *Clostridium botulinum* at 121 °C in saturated steam, but the process takes two hours at 160 °C in a dry air oven. During the first hour, the oven contents are warmed up to temperature after loading the oven, and the second hour of treatment is required to ensure that all the spores are killed.

Despite the disadvantage of the length of time dry heat takes to kill microbes, there are situations where its use is preferable to moist heat. Dry heat is used to sterilise glassware and metal instruments that would corrode if repeatedly exposed to moist heat. It is also used to sterilise heat-stable powders and water-repellent products such as petroleum jelly. Care must be taken with such products because heat penetrates them very slowly. Consequently they should only be sterilised in small batches with a large surface area to volume ratio.

Moist heat is used to sterilise many objects, although it cannot be used for all heat-resistant items because of the problem of corrosion. Although boiling water is hot enough to kill vegetative cells, it is not hot enough to kill spores. Boiling of drinking water does *not* sterilise it. It is, however, an excellent method of decontaminating water, rendering it safe to drink. The agents of water-borne infections are killed by boiling water.

To be effective, moist heat is employed at temperatures above boiling point and normal atmospheric pressures. This is achieved using an *autoclave*, a device analogous to the domestic pressure cooker. All the air within an autoclave is expelled by steam, which is then held in the autoclave chamber at a typical temperature of 121 °C and 15 pounds per square inch (lbf/in^2 ≈ 6.9 × 10^3 Pa) pressure. This is a sufficiently high temperature to kill almost all microbes, but the causative agents of scrapie and other spongiform encephalopathies are considerably more heat-resistant and require a temperature of 135 °C to inactivate them.

The time an autoclave load must be held at its sterilising temperature depends upon the nature of its contents. Adequate time must be allowed for the penetration of steam to the centre of a load. To assist this, loads should not be tightly packed. In microbiology laboratories, all waste is considered to be potentially infectious, and so everything to be disposed of is autoclaved prior to being discarded.

The efficacy of autoclaving may be assessed in different ways in addition to monitoring the temperature of the load during an autoclaving cycle. Spores of *Bacillus stearothermophilus* are very heat-resistant. Filter paper discs soaked in a spore suspension are often included in the centre of autoclave loads. After processing, these discs are used to inoculate broth cultures. The absence of growth after suitable incubation is indicative of a successful run. Unfortunately, this test involves an inevitable time delay whilst the culture is being incubated. To avoid this delay *Browne's tubes* may be included in an auto-clave run. These tubes contain a red chemical that turns green after exposure to sterilising conditions in an autoclave. *Bowie–Dick* autoclave tape is also used to indicate the effectiveness of heat sterilisation. It contains heat-sensitive strips that turn dark brown after sufficient heat exposure.

Some laboratory growth media, although reasonably heat-stable, are unable to withstand the rigours of autoclaving. This is particularly true for broths used in sugar fermentation tests, where the carbohydrate is prone to degradation if autoclaved. To render these fit for use, **Tyndallisation** is employed. On the day of preparation, the medium is held in a steam bath at a temperature between 90–100 °C for ten minutes. This is sufficient to kill the vegetative cells in the medium, but it does not kill spores. The medium is then held at room temperature for 24 hours. This time permits spores to germinate. The resultant vegetative cells are killed by steaming on the second day as on the first day. The whole process is repeated on the third day to remove any spores left after the second steaming. After three steamings, the medium is ready for use. Tyndallisation is too severe for egg-based media such as Dorset egg slopes or Lowenstein–Jensen medium, and these media require **inspissation**. This involves holding the media at 85 °C rather than near boiling point on each of three successive days. In every other respect Tyndallisation and inspis-sation are the same.

Foodstuffs such as milk are themselves sensitive to heating and even Tyndallisation causes unacceptable changes in the product. In order to reduce the microbial contamination to safe levels, **pasteurisation** is used. To pas-teurise milk it is held at 71.7 °C for at least 15 seconds. This is the high-temper-ature short-time (HTST) process. An earlier regime was low-temperature holding (LTH) and involved holding milk at 62.8 °C for at least 30 minutes.

LTH processing was sufficient to kill most milk-borne pathogens including *Mycobacterium tuberculosis*, which at one time was considered to be the most heat-resistant pathogen in milk. The discovery that *Coxiella burnetii*, the causative agent of **Q fever**, was capable of transmission in pasteurised milk, and also that it could survive LTH pasteurisation prompted the establishment of HTST processing, since this regime is effective against this organism.

Heat is not suitable as a means of sterilising the many objects or products that are heat-labile. Irradiation can prove to be an acceptable alternative in many instances. Ultraviolet irradiation is commonly used to sterilise work surfaces and air. At a wavelength of about 260 nanometres ultraviolet light is strongly absorbed by nucleic acids. It causes the formation of **pyrimidine dimers**. This leads to genetic damage and consequently to cell death. Unfortunately, ultraviolet light does not penetrate through glass or liquids, hence it is restricted to sterilising surfaces.

X-rays are more effective at causing the destruction of microbes than is ultraviolet light. However, they have little application in microbial sterilisation. X-rays are expensive to generate and are difficult to control. They have, however, been successfully used to generate genetic mutants for experimental studies.

γ-Irradiation, sometimes referred to as cold sterilisation, uses cobalt as ^{60}Co as a source of γ rays. Radioactive cobalt is a by-product of the nuclear industry, and thus it is a cheap source of γ rays. These rays penetrate materials efficiently, and γ-irradiation is used in the commercial sterilisation of antibiotics, hormones and other pharmaceutical preparations as well as medical plastic products such as catheters and syringes. γ-Irradiation has also found a controversial use in decontamination of foodstuffs, although this practice is not widespread.

Heat-labile solutions may be sterilised by filtration. Several filtration materials have been used, and all act to restrain microbes whilst letting solutions pass through into a previously sterilised collection vessel. Earthenware candles were popular as filters at one time, as were asbestos or sintered glass pads. However, the use of these items has been largely superseded by the development of cellulose filters. Originally nitrocellulose was used. This material has pores ranging in size from 3 micrometres to 10 nanometres, and probably worked as a result of adsorption as well as filtration. Filters may permit viruses to pass through, and so do not truly sterilise. Rather, they decontaminate solutions. Indeed, nitrocellulose filters have been used to determine the size of virus particles. Nitrocellulose filters have now themselves been superseded, and have been replaced by cellulose acetate membranes. These have a pore size of 0.22 micrometres or 0.45 micrometres.

Filtration is used to decontaminate heat-labile growth media and its supplements, pharmaceutical preparations and antibiotics. It is also used to sterilise air in laminar flow cabinets and certain types of operating theatre.

2.2.2   Chemical disinfection

Many and varied chemicals may be used as antiseptics or disinfectants. Antiseptics are used on living tissues, and so tend to be less toxic to humans than disinfectants. This was not always the case. Joseph Lister, when he was developing his theories of antisepsis during the latter part of the nineteenth century used highly toxic phenol preparations to prevent wound infections. The original purpose of surgical rubber gloves was to prevent surgeons from suffering phenol burns. The fact that it also reduced wound infection was an added bonus. This benefit has outlived the original purpose of wearing gloves in surgery.

Phenol was the first disinfectant in clinical use, and it is the standard by which other disinfectants are measured. It has a variety of effects on microbial cells, causing cell disruption, **denaturation** of proteins and loss of enzyme activity. It has marked anaesthetic properties, but it is also corrosive and highly toxic to humans. Many derivatives of phenol have been produced, and some are very effective germicides. Many derivatives lack the human toxicity exhibited by phenol. Phenol is bactericidal in a 1% solution, and is almost totally inactive at concentrations of 0.1%. A final concentration of 0.5% phenol has been used to preserve vaccines. Hexachlorophene is typically used as a 3% solution. Chlorhexidine is used as a 0.5–1% solution in isopropanol for skin disinfection or in aqueous solution for wound irrigation.

Alcohols such as methanol, ethanol and isopropanol are efficient killers of microbes. Absolute alcohols are less effective than 70% solutions in water. Alcohols act by disrupting membranes and coagulating proteins. Their ability to coagulate proteins may be demonstrated by cracking a raw egg into a dish of alcohol. Almost immediately the white of the egg takes on the appearance of being cooked. Bacterial spores and many viruses resist the effects of alcohol.

Chlorine and iodine act as powerful oxidising agents. These are rapidly germicidal, but are easily inactivated by the presence of organic matter. They are also irritant to humans. Chlorine compounds are added to swimming pools and, in lower concentrations, to municipal drinking water supplies. They are also widely used in the dairy and food industries. The most commonly used chlorine preparation is sodium hypochlorite or household bleach. This substance kills all known germs! Bleach typically contains 100 000 parts per million available chlorine, and is used as a 1% solution, giving 1000 parts per million available chlorine at its working strength.

Iodine is widely used as an antiseptic. Tincture of iodine comprises 2.5% iodine and 2.5% potassium iodide in 90% ethanol. Another popular preparation is 2% iodine in isopropanol. The irritation that iodine causes may be ameliorated when it is used as an **iodophore** preparation. Iodophores are complexes of iodine and an organic polymer. They permit a slow release of active iodine, extending the time of the antiseptic activity. Typically iodophores contain about 1% available iodine.

Two aldehydes, formaldehyde and glutaraldehyde, are commonly used as disinfectants. These react with proteins and nucleic acids to denature them. In this manner they exert their lethal effect. Both glutaraldehyde and formaldehyde are used in solution, but formaldehyde bombs are used to release formaldehyde vapour. Bombs are used, for example, to disinfect rooms heavily contaminated with microbes. Another important use for formaldehyde bombs is to decontaminate centrifuges in which potentially infectious samples have leaked.

Organically substituted ammonium compounds such as cetrimide and benzalkonium chloride are known as quaternary ammonium compounds. These act as cationic detergents, widely used in hospitals. They act by disrupting bacterial membranes and allowing leakage of the cell contents. They are inactive against spores. Certain disinfectants that are sold for domestic use are based on quaternary ammonium compounds. One of the appeals of such products is that they smell pleasant, but bacteria of the genus *Pseudomonas* are able to metabolise quaternary ammonium compounds rather than being killed by them. Indeed, they can use them as nitrogen and carbon sources. Cetrimide is therefore added to selective media used for the isolation of pseudomonads.

Chloroform vapour is used to kill vegetative bacterial cells and some viruses. It acts as an organic solvent to disrupt biological membranes, but it does not generally affect proteins. Chloroform is used to kill bacteriocin-producing strains growing on solid media, without affecting the activity of the **bacteriocin**. After exposure to chloroform, the producer strain is removed from the plate, and after the chloroform vapour has been given time to disperse, its position is cross-streaked with strains to be tested for bacteriocin sensitivity. Glass Petri dishes must be used for this test, because plastic is soluble in chloroform.

Few gases have marked antimicrobial activities, but ethylene oxide is an exception. In air, ethylene oxide is highly flammable, and so in practice it is mixed with 90% carbon dioxide to prevent it from burning. It is a highly effective disinfectant, capable of killing bacterial spores rapidly. Its activity is enhanced by the presence of moisture and moderately elevated temperatures. It is used to sterilise bulky items or delicate instruments that cannot withstand other types of sterilisation.

For many years, compounds containing heavy metals such as copper, mercury, zinc and even arsenic were used as antiseptics. These compounds are now considered to be too toxic for routine human use. However, silver salts are still used today. Silver nitrate drops are instilled into the eyes of newborn babies to protect against infections with *Neisseria gonorrhoeae* and *Chlamydia trachomatis* that may have been acquired from an infected mother during birth. Silver sulphadiazine is applied topically to help to prevent infection in burns victims. Heavy metal ions were once used in the treatment of **syphilis**. It used to be said that a night with Venus led to a lifetime with Mercury.

Acridine dyes are bactericidal due to their interaction with nucleic acids. They are used topically as antiseptics in the prophylactic treatment of mild burns. Other dyes have bactericidal properties, but these may be selective in their action. Species of the genus *Brucella* may be identified by their resistance or susceptibility to particular dyes (Fig. 2.5).

2.3   Microscopy

Microorganisms are, by definition, too small for individual organisms to be seen by the naked eye. Many microbes can grow together on solid media to produce visible colonies, some of which may be very large. In liquid culture, microbes may multiply to a sufficient density to make the culture appear turbid. In spite of these observations, in order to visualise a single microorganism its image must be magnified using a microscope. Today, compound light microscopes are routinely used to visualise microorganisms other than viruses. Compound light microscopes are no more than optical benches set up in a manner that makes them comfortable for humans to use. Observation of microscopic preparations can be made from a position that suits human posture, rather than the demands of an ordinary optical bench. They comprise three units: an illumination system, an object and an analytical system. The *object* to be viewed is mounted on a glass microscope slide that is held on the microscope stage in the *object plane*. Light is focused on the object by means of the *substage condenser*. The image of the object is magnified by the combined action of two lens systems situated above the object. These are the *objective lens* and the *eyepiece* or *ocular lens*. Together, the objective and ocular lenses comprise the *analytical system* of the microscope (Fig. 2.6).

Commonly, eyepieces magnify 10×, and the objectives multiply 10×, 40×

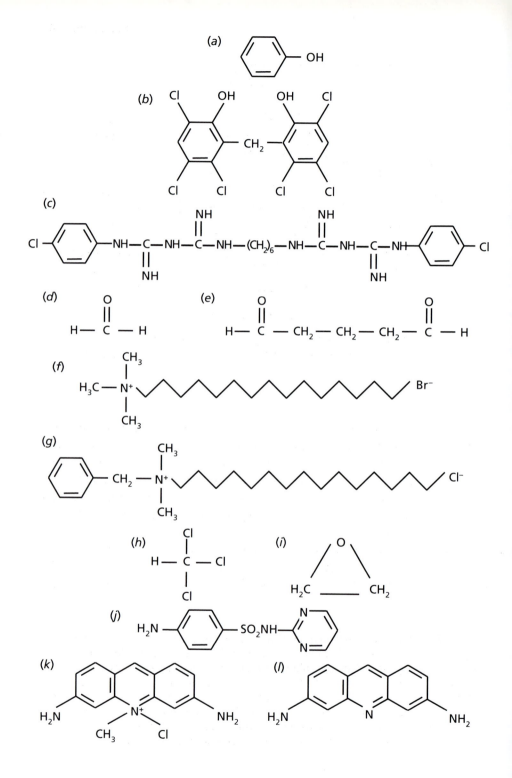

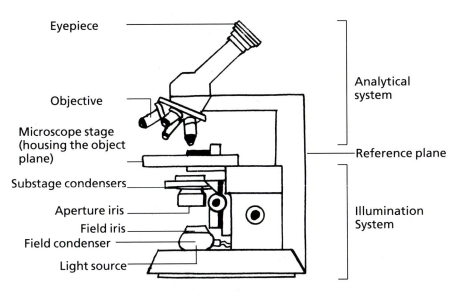

Fig. 2.6. A stylised view of a compound light microscope. In many light microscopes, the light source is held in the base of the instrument, rather than as illustrated.

and 100×. The magnification of an object is obtained by multiplying the magnification of the eyepiece lens by the magnification of the objective lens. Thus objects are magnified 100×, 400× and 1000× by commonly used combinations. Useful magnification is limited, however, by the phenomenon of *resolution*. The resolving power of a microscope is its ability to permit objects located close to one another to be distinguished as two separate and distinct entities. The limit of resolution is the smallest distance at which two points appear as separate entities (Fig. 2.7). The resolving power of a microscope is governed by the wavelength of light used to illuminate the object, and also the *numerical aperture* of the objective lens. The numerical aperture of a lens is a

Fig. 2.5. (Left) Chemical structures of selected antimicrobial agents. (*a*) Phenol; (*b*) Hexachlorophene; (*c*) Chlorhexidine; (*d*) Formaldehyde; (*e*) Glutaraldehyde; (*f*) Cetrimide; (*g*) Benzalkonium chloride; (*h*) Chloroform; (*i*) Ethylene oxide; (*j*) Sulphadiazine; Acriflavine, an acridine dye that is a mixture of two components, (*k*) euflavine and (*l*) proflavine. The hydrocarbon side-chain of cetrimide is of variable length, being 12, 14 or 16 carbon atoms long. The hydrocarbon side-chain of benzalkonium chloride is also variable, ranging from 8 to 18 carbon atoms in length. Silver sulphadiazine is used topically to help protect burns victims from infection.

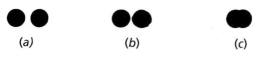

(a)   (b)   (c)

Fig. 2.7. Resolution of objects. The first pair of objects (a) are easily resolved as two distinct individual items, the objects labelled (b) are just resolvable, and the pair of objects marked (c) cannot be distinguished as separate entities. The distance between the middle pair of objects is thus approaching the resolution of the figure.

measure of its light-gathering capacity. This number is engraved on the objective together with its magnifying power.

The relationship between the resolution of a microscope (d), the wavelength of light (λ) and the numerical aperture (NA) is expressed in the following formula:

$$d = \frac{0.61\lambda}{NA}$$

It can be seen that the greatest resolution is obtained using light of the shortest wavelength possible, and a lens with the maximum numerical aperture. The numerical aperture of a lens is defined by the following equation:

$$NA = n \sin \theta$$

Here, n is the refractive index of the medium between the object and the objective, and θ is half the angle of the cone of light entering the objective from the centre point of the object. In some texts μ represents the refractive index of the medium. However, this may cause confusion for students of microbiology since μ is also used to represent the specific growth rate. Here, n is used to represent refractive index. This is to avoid confusion over the use of the symbol μ. When using air as the medium, the maximum theoretically achievable numerical aperture for an objective is 1, since the refractive index of air is 1 and the maximum angle for the cone of light entering the objective is 180°. Half of 180° is 90°, and sin 90° = 1. In practice, the maximum achievable numerical aperture for a dry objective is no more than 0.85. The wavelength of green light is 0.55 micrometres, and if these values are substituted in the formula above, the limit of resolution can be determined:

$$d = \frac{0.55}{2 \times 0.85} \quad \text{micrometres}$$

therefore: $d \approx 0.3$ micrometres

The resolution power of a microscope can be enhanced considerably by placing oil between the object and the objective. Cedar oil has a refractive index of 1.52. This was the oil that was traditionally used for oil-immersion, but it has been superseded by synthetic oils: the numerical aperture of oil-immersion lenses is typically about 1.3. This gives a resolution of about 0.2 micrometres when objects are viewed using oil-immersion. Most bacterial preparations are viewed under oil-immersion to improve the resolution of the image. However, oil-immersion can only be used in conjunction with objectives that are designed for the purpose. Objectives for use with oil-immersion are appropriately engraved.

Microbes can be visualised using a variety of microscopic techniques, and different types of microscopy are used for different purposes. All the techniques rely, however, on the fact that microorganisms interfere with the path of light or electron beams. These interference effects can be exploited in different ways. It would be very expensive to have to stock a laboratory with light microscopes dedicated to a particular technique; one for bright-field microscopy, one for dark-ground microscopy, etc. To overcome this problem, modern compound microscopes have a modular design. It is possible to replace individual components such as objectives or the substage condenser to build up a customised system for each particular function.

2.3.1 Bright-field microscopy

The most commonly used light microscope system in microbiology is bright-field microscopy. This technique, however, permits only the examination of dead microorganisms. The entire field is evenly illuminated when using the Köhler illumination system. This involves the use of an extra substage condenser lens that allows light to fill the optical field as illustrated in Fig. 2.8. Microbes appear as dark objects against a bright background. Since many bacteria do not produce pigments, stains are used to enhance their visibility in bright-field microscopy.

Microscopes require to be set up to optimise the light path through the instrument, and thereby to obtain the best image of the preparation. Many students feel very daunted when first using compound microscopes, but the steps involved in setting up a microscope for bright-field illumination are relatively simple, and they do have a logical basis. Many modern compound microscopes have an integral lamp that is regulated by a voltage control unit. The lamp acts as the source of illumination, and light from the lamp passes through a *field condenser* and a *field iris*. This regulates the area of the field of illumination. Above the field iris lies the *aperture iris*, also known as the *substage condenser iris*. This iris regulates the angle of the cone of light leaving the

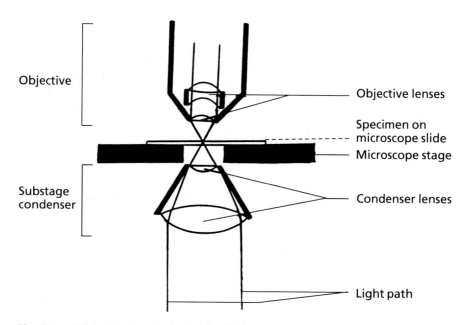

Fig. 2.8.   Köhler illumination for bright-field microscopy.

condenser and consequently entering the objective. This, in turn, controls the resolution of the instrument.

When setting up a compound microscope for bright-field microscopy, both the field iris and the aperture iris are fully opened. Using the objective with lowest magnification, an object is placed on the microscope stage and is brought into focus. Even though microbes are viewed at much higher magnification, the reason an object is focused using the lowest power objective first is to optimise the conditions for correcting misalignment of the optical path. As the magnification increases, so too does the apparent misalignment, and if the optical path is badly aligned, it is impossible to correct using objectives of high magnification. The next step is to ensure that light from the lamp is focused on the object. To do this, the field iris is closed and the substage condenser is raised and lowered until the edge of the field iris is brought into sharp focus. The light source must then be centred so that the light beam can pass directly through the object, and then up through the analytical system along the optical axis. When the light on the object is focused and centred the aperture iris is then adjusted to give the maximum resolution. Bright-field microscopy is most frequently used in conjunction with stained preparations, for example to determine whether a bacterium is Gram-positive or Gram-negative, or in a Ziehl Neelsen stained preparation used in the diagnosis of **tuberculosis**.

2.3.2   Dark-ground microscopy

Some microorganisms have structures that are too fine to be visualised using bright-field microscopy. *Treponema pallidum*, the bacterium that causes syphilis, is a slender spirochaete that is too delicate to be seen under bright-field illumination. Exudate from a syphilitic lesion contains live bacteria, and these may be visualised using dark-ground microscopy, thus providing a provisional diagnosis of syphilis. Dark-ground microscopy is frequently used to visualise live organisms, and is especially useful in observing motility. The theory behind dark-ground microscopy is that particles in a light beam cause the light to become dispersed. An opaque plate is placed in the substage condenser so that light from the lamp cannot enter the objective directly. Rather, this plate causes the object to be illuminated by a hollow cone of light that passes just outside the edges of the objective lens system. If there is no object present in the object plane, the field of view appears black. However, if a specimen is placed in the object plane, this will cause some light to be dispersed. The dispersed light is able to enter the objective, causing a brilliantly illuminated image of the specimen to appear against the dark background (Fig. 2.9).

2.3.3   Phase contrast microscopy

Although microorganisms have very simple structures, they are composed of components whose light-refractive properties vary from their surrounding medium. That means that a light beam passing through a microbial cell will be bent and retarded to different extents as it passes through various structures and will exit in a different wave-phase than light that has not passed through the microbe. This phase difference occurs in bright-field illumination; but because of the intensity of the background lighting, it is far too weak to be exploited usefully when the microscope is properly set up. Phase contrast microscopes are set up to enable microscopists to exploit phase differences. These become translated into differences in light intensity when light passes through a specimen. Differences in phase may result in *constructive interference* when the object appears brighter than its surroundings, or *destructive interference*, leading to the production of a darker image. As the light leaves the object the phase differences cannot be detected by the human eye.

As with dark-ground microscopy, the object is illuminated with a hollow cone of light created by an annular stop in the substage condenser. In phase contrast microscopy, however, the hollow cone of light enters the objective lens system. If the light is not diffracted, it passes through a phase plate that

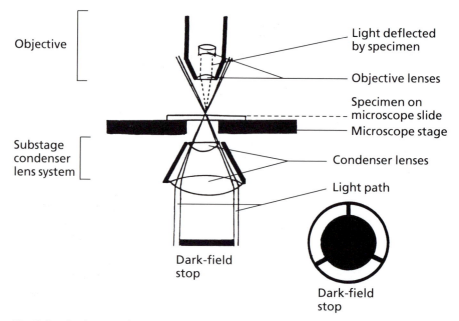

Fig. 2.9. Dark-ground microscopy. When using low power objectives, a dark-field stop can be placed below the substage condenser of a normal compound microscope, but if high power objectives are used, a special dark-ground condenser must be employed, with an integral dark-field stop fitted.

has a phase ring etched into it in a position to accommodate precisely the cone of direct light. The depth of the ring is designed to retard direct light rays by one quarter of a wavelength as they pass through the ring. Diffracted light rays pass through the phase plate unchanged. Light from the object suffers interference as a result of its passage through the phase plate, interference occurring as light leaves the phase plate. This interference may amplify the light, in which case the object will appear *phase bright*, or may cause a diminution of the signal with the image appearing *phase dark*. Consequently an enhanced phase contrast image of the object is produced. The phase plate enhances the phase contrast of the specimen enabling phase differences to be visualised easily. Each phase contrast objective has to have its own annular stop in the substage condenser to be effective (Fig. 2.10).

Phase contrast microscopy has a great advantage over other optical systems in the examination of microbes. It relies simply on the physical structure of the specimen, and thus living cells may be viewed with ease. Phase contrast microscopy has been of great value in the study of dynamic cellular processes

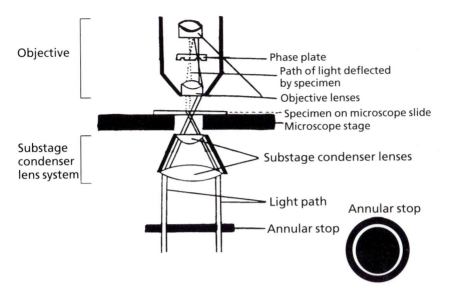

Fig. 2.10.   Phase contrast microscopy. Solid lines show the path of light that is not refracted. The dotted lines show the path of light that has been refracted by the object being viewed. The phase plate has a ring inscribed into its surface to match precisely the ring of the annular stop located in the substage condenser. It is this ring that retards light by a quarter of a wavelength, allowing the image to be enhanced.

such as bacterial spore formation and germination. It avoids the need to examine killed and fixed specimens.

2.3.4   Fluorescence microscopy

Fluorochromes are chemicals that absorb light of short wavelength, and emit light of a longer wavelength. This will result in a change in the colour of the light emitted by a fluorochrome. If ultraviolet light is absorbed, then visible light is emitted in the process of fluorescence. Fluorochromes may be used as fluorescent dyes to stain microorganisms. When fluorescently stained, microbes visualised in fluorescence microscopes appear as brightly coloured cells against a dark background. This makes the detection of scanty cells in a stained preparation much easier than when using bright-field microscopy. The detection of mycobacteria stained with auramine has greatly enhanced the ability to make a provisional diagnosis of tuberculosis when only scanty mycobacteria are present.

In addition to fluorochromes being used directly, they may be attached to

antibodies. Fluorescently tagged antibodies may then be used to flood a microscope preparation to locate the position of **antigens**. This technique is widely used to locate viruses in infected tissue specimens using antibodies raised against virus antigens. Chlamydial infections are also diagnosed in this manner, since chlamydia can be cultured only with difficulty in mammalian tissue cultures. Furthermore, antibodies from one animal can be used as antigens to raise antibodies in another animal, and these antibodies can be fluorescently labelled. This technique is exploited in the **serological diagnosis** of syphilis. Cells of *Treponema pallidum* are fixed to a slide and this is flooded with **serum** taken from the patient. If the patient has syphilis, the serum will contain antibodies that will attach to the fixed *Treponema pallidum* cells. After the serum is washed off, any anti-treponemal antibodies will remain on the slide, attached to the bacterial cells. These antibodies may be detected using anti-human antiserum in which the anti-human antibodies have been fluorescently labelled. If the patient has syphilis the treponemes will appear fluorescent when viewed in a fluorescence microscope, but if the patient does not have the disease, then the anti-human antibodies will have nothing to attach to, and the preparation will appear dark.

Fluorochrome dyes are not necessarily used to dye whole cells or to tag antibodies. Di-amidino-phenyl-indole (DAPI) is a fluorochrome dye that is used to determine the intracellular location of DNA. It has been used in the study of yeasts that have lost their mitochondrial DNA to demonstrate the absence of this material.

Older fluorescence microscopes were transmission instruments, set up in a fashion similar to that of bright-field microscopes, but with a short-wave rather than a white light source. Modern instruments employ incident light, and the microscope objective also serves as its own condenser. Modern fluorescence microscopes employ a short-wave light source at right angles to the optical axis of the instrument. The light beam is directed onto the specimen by means of a **dichroic** mirror placed at 45° to the beam of light and within the optical axis of the instrument. This will reflect the short-wave light onto the specimen where it may excite any fluorochrome present to emit long-wave light. The dichroic mirror will allow the long-wave fluorescent light to pass through to the eyepiece where it may be detected. Before entering the eyepiece, light from the specimen passes through a barrier filter that blocks short-wave light, allowing only long-wave light to be detected (Fig. 2.11).

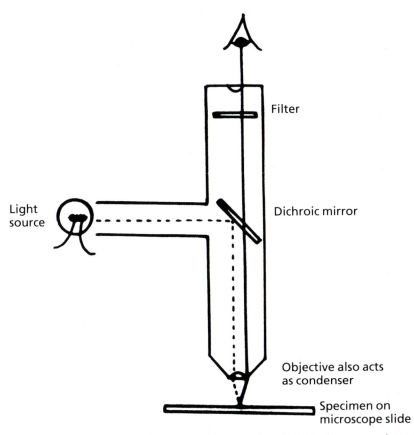

Fig. 2.11.   Fluorescence microscopy. Short wavelength light, shown as a dotted line, is directed from a light source through the objective onto the object, where it excites fluorochrome dyes to emit visible light, shown as a solid line. This passes back up the objective and through a filter to cut out any reflected short-wave light before it is observed through the eyepiece.

2.3.5   Electron microscopy

Transmission electron microscopy was the first system of electron microscopy to be developed. It operates on theoretical principles identical to those exploited in light microscopy, and is used to visualise objects that are too small to be seen using light microscopes. Viruses fall into this category because of the limitations of resolution imposed by the wavelength of visible light. This restriction is overcome if an electron beam is used in place of visible light, since electron beams have an extremely short wavelength. Using

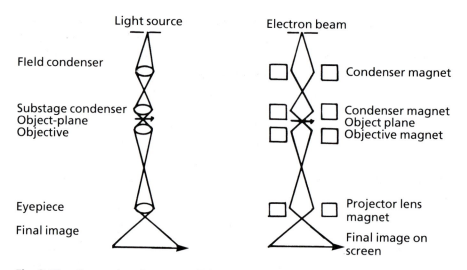

Fig. 2.12. Comparison between a light microscope and a transmission electron microscope.

electron microscopy it is possible to resolve points only 5 angstrom units apart. This is equivalent to a distance of 0.5 nanometres. In electron microscopy, magnets are used to focus the electron beam rather than the glass lenses used in light microscopy, and the dyes used are electron-dense rather than coloured compounds. Heavy metal ions are used to absorb electrons in electron microscopic preparations. Typically these are provided in the form of lead citrate or uranyl acetate. Osmium tetroxide is used in electron microscopy as a fixative and the osmium ions also serve as dye since they are electron-dense.

Transmission electron microscopes are in many respects analogous to light microscopes (Fig. 2.12). They require the electron beam to pass through a vacuum. If this is not done, molecules in the air deflect the electron beam, and a sharp image cannot be obtained. Extremely thin specimens must be used for transmission electron microscope studies. These may be prepared in a number of different ways. Although transmission electron microscopy produces stunning and often very beautiful pictures, the techniques of specimen preparation employ very harsh conditions, and this may cause considerable damage to specimens. Consequently, the images produced are prone to artefacts, and these make the interpretation of electron microscope images very difficult.

In *ultrathin sectioning*, staining samples are embedded in a plastic resin and cut into extremely fine sections using a specially prepared glass or diamond knife in an apparatus known as an ultramicrotome. These ultrathin sections are then mounted on a copper grid prior to examination. Ultrathin sections allow the internal structures of cells to be examined. Ferritin-labelled antibodies may be used to detect specific antigens in ultrathin sections in a process analogous to immunofluorescent microscopy.

Negative staining exploits the electron dense properties of compounds such as phosphotungstic acid. The heavy metal ions do not penetrate the specimen, but form thick deposits upon the background material and in crevices within the specimen, enhancing the contrast between these structures and the object to be visualised. This method is routinely used for the detection of viruses in clinical specimens and is useful for studying the external appendages of bacteria, e.g. flagella.

Shadowing is a technique used to display the surface topology of a structure. Specimens are mounted on a grid and shadowed with a thin layer of a metal such as platinum or gold. The shower of metal particles is directed at the specimen at an oblique angle, so that only the surface facing the shower will be coated. This causes a shadow to be cast in the lee of the particles. This technique is used to study the external structures of viruses. The topology of plasmid DNA is also studied using shadowing. DNA preparations are first coated in a layer of protein to make the molecules easier to visualise. The cost of the metals used for this preparation limits its routine use.

In *freeze-etching*, rapidly frozen samples have their superficial layers fractured. A carbon replica is produced of the exposed sample surface, and the replica is then examined. Fracturing occurs along the weakest planes within a structure, and frequently along the surface of membranes. Freeze-etching is used to study the internal structures of microbes. Because flash freezing is employed, the internal structures are not exposed to the distortions caused by other fixative methods, but the results are often more difficult to interpret.

Transmission electron microscopy is similar in many respects to light microscopy. A more recent development is that of *scanning electron microscopy*. An electron beam is used to scan the surface of a specimen in a manner somewhat similar to the scanning of a television tube. This causes the surface of the specimen to emit secondary electrons and other types of radiation at various angles, producing a three-dimensional image. The emitted radiation is detected in a scintillator, and the signal is amplified to produce an image on a cathode ray tube or other recording device. The closer a surface is to the electron beam, the more secondary radiation it will emit, and the brighter will be its final image. Scanning electron microscopy is used to produce striking images of the surfaces of structures, although it does lack the very fine resolution associated with transmission electron microscopy.

2.4   Simple and differential staining of bacteria

Most bacteria do not produce pigments and they are also poorly refractile. This means that using bright-field microscopes they can be difficult to visualise. To overcome this problem, bacterial cells may be stained with a variety of coloured dyes. These enhance the contrast between the cells and their background. *Simple stains* dye cells or their background with a single colour. There are numerous compounds that can be used as simple stains. These include methylene blue and malachite green, which stain blue and green, respectively. Red dyes commonly used as simple stains include dilute carbol fuchsin and safranine. Simple stains may also act as *negative stains* where the background rather than the bacterial cell is stained. The Indian ink capsular stain provides a good example of a simple negative stain. The ink particles cannot penetrate the capsular material, and these structures appear as clear haloes around bacterial cells suspended in a dark medium. The red dye eosin can also be used as a negative stain to visualise capsules. *Differential stains* are more complex, and involve the exploitation of a combination of dyes with different properties. Differential stains may be used to distinguish different cell types, or to visualise particular cell structures.

2.4.1   Gram staining

The Gram stain, first described by Christian Gram in 1884, was developed in an attempt to differentiate bacteria within tissue sections. The basis of the staining protocol is that certain bacteria retain a blue-black dye complex when exposed to organic solvents such as acetone or alcohol. The complex is formed between iodine and an aniline dye such as crystal violet. Tissues do not retain this complex, and they require counterstaining with a red dye to assist visualisation. It was soon realised that not all bacteria retain the crystal violet–iodine complex when decolorised with acetone or alcohol. Those bacteria that do retain the complex are known as **Gram-positive**, and those that are decolorised by acetone or alcohol are **Gram-negative**. This reaction demonstrates a fundamental difference in the structure of Gram-positive and Gram-negative bacteria, particularly with respect to their cell wall structure, and this forms the basis of many bacterial identification schemes.

Despite the fact that it is over 100 years since the Gram stain was first described, its precise mechanism is still not fully understood. However, Gram-positive bacteria have cell walls that have a structure radically different from that of the cell walls of Gram-negative bacteria. Gram-positive cell walls are composed primarily of many layers of peptidoglycan, and this is thought to help to retain the crystal violet–iodine complex. In contrast, the cell walls of Gram-negative bacteria contain only one or two layers of peptidoglycan, and this is insufficient to retain the crystal violet–iodine complex. This explanation of the Gram reaction is not completely satisfactory. There are bacteria that have cell wall structures that resemble those of Gram-positive bacteria, but after Gram staining they appear as Gram-negative cells. Similarly, certain archaebacteria that entirely lack peptidoglycan in their cell walls appear Gram-positive after Gram staining. Although most bacteria can be classified as Gram-positive or Gram-negative, some bacteria display an intermediate reaction with the Gram stain. Part of a cell may appear Gram-positive and the remainder Gram-negative. Such bacteria are termed **Gram-variable**. A further difficulty may arise when examining Gram-stained films of bacteria. Some bacteria appear Gram-positive when examined in fresh cultures, but lose the ability to retain the Gram stain when older cultures are examined.

The protocol for Gram-staining bacteria is as follows. A fine suspension of bacteria is spread on the surface of a clean microscope slide and allowed to dry slowly in the air. The cells are then fixed to the glass by gently passing the slide two or three times through a hot Bunsen burner flame. The film is then flooded with crystal violet. This will colour both Gram-positive *and* Gram-negative cells. After about 30 seconds, the crystal violet solution is washed off, and replaced with a dilute iodine solution. The iodine forms a blue-black complex with the crystal violet. At this stage Gram-positive and Gram-negative cells are still indistinguishable. After about 1 minute, the iodine solution is washed off from the film. The slide is then decolorised by flooding with an organic solvent such as acetone or alcohol. It is at this stage that Gram-negative cells lose their colour and this is the critical step in the Gram-staining procedure. Acetone is the solvent of choice, and normally it requires only 2 or 3 seconds to react before being washed off. If acetone is left on a film for too long, the crystal violet–iodine complex will leach out from Gram-positive cells as well as Gram-negative cells. Conversely, if the acetone is not applied for long enough, Gram-negative cells will not become decolorised, and will appear Gram-positive in the final preparation. After acetone treatment Gram-negative cells will appear colourless, and therefore the film requires counter-staining with a red dye such as dilute carbol fuchsin. The Gram-staining procedure causes a degree of shrinkage of bacterial cells, partly as a result of the loss of water. Consequently, after Gram-staining bacteria appear to be about two thirds of their true size.

2.4.2   Ziehl Neelsen staining

Mycobacteria have very unusual cell walls. They contain considerable amounts of mycolic acid, a waxy substance that makes their cells difficult to stain using conventional stains. They may, however, be stained using the Ziehl Neelsen procedure. A heat-fixed film is prepared and flooded with strong carbol fuchsin. The slide is then heated until the carbol fuchsin solution starts to steam. The dye solution should not be allowed to boil or dry out, but should be kept steaming for 5 minutes. The heat allows the dye to penetrate through the waxy cell wall. The film is then washed with filtered water. It is important to use filtered water because tap water may contain mycobacteria and these may give a false positive result. Carbol fuchsin is a basic dye and reacts with acid to produce a yellowish-brown compound that may be leached easily from tissues. The leaching process is enhanced by alcohol. However, the waxy walls of mycobacteria protect the dye within their cells when exposed to a solution of 3% hydrochloric acid in 95% alcohol. When films are treated with acid-alcohol, only mycobacteria retain the carbol fuchsin and hence appear red. Methylene blue is generally used as a counterstain to colour other material in the film, although some people prefer to use malachite green as a counterstain. Because of their behaviour, mycobacteria are referred to as **acid-alcohol fast** bacteria. Other bacteria, such as those in the genus *Nocardia*, can retain carbol fuchsin when decolorisation with 5% sulphuric acid is used, but are themselves decolorised when acid-alcohol is employed. These are known as **acid-fast** bacteria.

Bacterial spores may be stained using a modification of the Ziehl Neelsen method. Films are stained in the usual manner, but are decolorised for only 30 seconds to 2 minutes with 0.25% sulphuric acid. The exact timing of the decolorisation step is determined experimentally. Alternatively, decolorisation may be accomplished by dipping the slide into a solution of 2% nitric acid in absolute alcohol, and then washing it immediately with plenty of water. Methylene blue is used as a counterstain to stain the vegetative cells blue whilst the spores appear red because they retain the carbol fuchsin.

2.4.3   Other differential stains

In addition to distinguishing cell types, differential stains may be used to reveal bacterial inclusions. Bacteria such as *Corynebacterium diphtheriae*, which causes diphtheria, contain **volutin granules**. These are composed of polyphosphates, and used as a chemical energy store. Volutin granules are sometimes called **metachromatic granules** because of their interaction with certain dyes. Volutin granules may be stained using the Albert–Leybourne method. Albert's original stain used a mixture of toluidine blue and methyl green in glacial acetic acid and ethanol. Leybourne substituted malachite green for methyl green since this enhances the colour of the final preparation. Films are flooded in Albert's stains for 3 to 5 minutes, then washed and dried. The preparations are then flooded with a dilute solution of iodine for 1 minute before washing, drying and examination. Volutin granules appear blue-black, and the cytoplasm looks greenish-blue when stained using this method.

Sudan black is used to stain intracellular lipid. The unincorporated dye is removed from the preparation with an organic solvent, and the protoplasm is counterstained with either safranine or dilute carbol fuchsin. Lipid appears blue-black and the bacterial protoplasm appears pink.

Other bacterial structures may also be differentially stained. Cell walls may be stained using Hale's method. A rather thick inoculum is placed on a cover-slip and is treated without fixation with a 1% solution of phosphomolybdic acid. This acts as a **mordant** allowing the dye to become firmly fixed to the cell wall material. The coverslip is carefully washed, and then floated in a fresh 1% aqueous solution of methyl green. The coverslip is then mounted, film down-wards, on a drop of sterile distilled water, and can be sealed with melted petro-leum jelly, to avoid evaporation.

The location of DNA in cells may be observed using Robinow's stain. This is best performed on young, actively growing cultures, and this is conveniently achieved by harvesting cells that have been growing on nutrient agar for 3 hours. The agar culture is cut into blocks, and these are fixed using osmium tetroxide vapour. A clean coverslip is then placed on the block to obtain an *impression preparation*. This is then exposed to secondary fixation using mer-curic chloride in 30% ethanol. Any RNA present in the cells is then hydrolysed by treatment with a 1 molar solution of hydrochloric acid at 60 °C. Finally, the remaining DNA is stained with Giemsa stain, and the coverslip may be mounted on a drop of sterile distilled water.

2.4.4   Visualising fungi

In contrast to bacteria, heat-fixed smears of fungi are generally useless for microscopic examination. It is thus necessary to examine either stained or unstained wet mounts of fungi. Furthermore, because of their size, it is gener-ally unnecessary to use an oil-immersion objective to view preparations of fungi.

Use of stains such as Gram's stain and a modification of the Ziehl Neelsen stain, used principally in bacteriology, is somewhat limited in mycology. Gram's stain can be used to visualise yeasts but the results are often unsatisfactory. In clinical samples that have been stained with Gram's stain, individual yeast cells generally appear as **Gram-variable**, but with large areas of the cell retaining the crystal violet–iodine complex. Methylene blue can also be used to stain yeasts but lactophenol cotton blue, a combination of a stain and a mountant, is the preparation of choice for the microscopic examination of yeasts. Fungal spores can be stained using a modification of the Ziehl Neelsen stain, developed to stain mycobacteria. Strong carbol fuchsin is introduced into the spores by heating a fixed preparation, and the stain is removed from other structures by treatment with 5% sulphuric acid. Methylene blue or malachite green are used as counterstains. Spores stain red, and other structures take up the colour of the counterstain. Ascospores can be demonstrated in this manner.

The most common method used to visualise fungi is to use lactophenol cotton blue. Light green is an alternative stain to cotton blue that can be used to stain fungal structures. However, it does not stain fungi so intensely, and it tends to fade with time. It is therefore not suitable for preparing mounts that are to be kept for long periods. Cotton blue stains hyphae blue, particularly young hyphae, and the intensity of staining increases with time. Old mounts can become overstained. Lactophenol is prepared by dissolving phenol crystals, lactic acid and glycerol in distilled water. When it is used on its own as a mountant, lactophenol can be employed to observe the natural colours of fungal structures.

To visualise fungi microscopically, a portion of the fungal growth is removed from the culture using a sterile needle or wire hook. The sample is then placed in a drop of alcohol on the surface of a clean microscope slide. The alcohol is used to drive out air that becomes trapped between the fungal structures. A drop of lactophenol cotton blue is then added to the preparation. Yeast cultures are generally emulsified directly in lactophenol cotton blue and the culture material is more easily manipulated with a wire loop. Care should be taken not to add too much mountant to the preparation, and if this accidentally happens, then the excess should be gently removed using blotting paper. The edge of a coverslip is then placed onto the drop of mountant carefully and the coverslip is lowered over the preparation, avoiding the introduction of air bubbles. Some authorities recommend that the preparation is then gently heated to enhance staining, but this is generally unnecessary and may cause overstaining unless extreme caution is applied.

With certain fungi it can be difficult to remove sufficient material from the culture. This is particularly the case with tough, compact colonies.

Alternatively, powdery colonies may yield only spores. In such cases it is often advantageous to remove a little of the agar substrate along with the fungal material. With any preparation from a mould culture, it is usually necessary to tease out the fungal growth on the slide to make visualisation easier. This is generally done in the alcohol drop with two sterile mounted needles. It is often necessary to add more alcohol to prevent the fungus from drying out during this procedure.

Mounts prepared in lactophenol can be kept for several weeks or months. More permanent preparations can be made by sealing the edges of the coverslip. This is achieved by painting on several layers of clear nail varnish around the edge of the coverslip. Each layer of nail varnish must be allowed to dry before application of the next layer. If this is to be successful, then it is important that the edges of the coverslip are clean and dry before applying the nail varnish. Properly sealed, such preparations can be kept for many years.

2.5 Enumeration of microbes

There are many occasions when it becomes necessary to count the number of microorganisms in a certain location or within a specimen. In monitoring the microbial quality of drinking water, assessing the level of bacterial contamination in an operating theatre and in quality control in the food and pharmaceutical industries, the numbers of microbes present is of prime importance. The method used to count microorganisms depends upon the type of information required, the numbers and types of organisms present, and the physical nature of the sample. Counting microbes can be accomplished in several ways. Some methods yield a *total cell count* taking no account of whether the microbes are viable or not. Alternatively, *viable cell counts* are used to determine the numbers of organisms capable of propagating themselves. There is much current interest in improving our ability to count microbes. This includes developing techniques to count the total microbial load and the number of live organisms present within a sample. However, some microbes remain metabolically active, but in a non-cultivable form. These pose a special problem for scientists trying to enumerate them.

2.5.1 Total cell counts

A variety of techniques may be employed to determine the total microbial count of a sample. These include direct microscopical examination, measurement of the turbidity of a suspension, determination of the wet or dry weight of a culture, examination of the chemical composition of a sample or by measurement of electrical impedance.

Microscopic counts of microbial cells   Microorganisms may be counted directly by observing cells in microscopic preparations. Typically this is done by placing a suspension in a counting chamber. These are specially prepared glass slides with an accurately measured grid etched onto the middle section of the slide. The two outer sections of the slide are raised 0.05 millimetres above the central wall. When a coverslip is placed over the suspension, this fixes the volume of the sample over each square in the grid. The mean number of microbes lying over each square of grid is calculated by direct counting, and since the volume of suspension over each square is fixed, the number of organisms in the original sample can be calculated.

Counting microbes using counting chambers may be difficult if the organism being counted is motile. This difficulty may be overcome by diluting the sample with formol saline. This kills microorganisms in the suspension, rendering them non-motile and therefore easier to count. Phase contrast microscopy is frequently used to examine the counting chamber to improve the visualisation of unstained bacteria.

The *Breed count*, used to assess the quality of milk, is a variation on this method. This is an old technique that fell out of favour for a considerable time. However, it is a cheap technique that is rapidly carried out and it does yield valuable information about the milk sample and, indirectly, about the herd from which the milk was drawn. Because of this, it is now undergoing a renaissance. Precisely 10 microlitres of milk is taken up in a Breed pipette, and then delivered onto a 1 centimetre square etched onto a clean microscope slide. This milk is then spread over the square with a straight inoculating wire. The film is air dried and stained with Newman's stain. This fixes, defats and stains the milk preparation in a single operation. After drying, the Breed film is then examined microscopically using a 100× objective. The number of bacteria and **leukocytes** in ten randomly selected fields are counted. From this, the average number of bacteria and leukocytes per high power field may be determined. There are 3000 high power fields per square centimetre. Since 10 microlitres of milk was spread over this area, the number of bacteria and leukocytes present in the original sample can be calculated by multiplying the average count per high power field by 3×10^5 per millilitre.

The Breed count gives the total number of bacteria in a sample, irrespective of their viability. Hence it cannot be used to assess the effectiveness of pasteurisation in reducing the viable count of potential pathogens. It does, however, rapidly give an indication of the number of the bacteria and leuko-cytes in milk. High leukocyte counts are indicative of disease such as **mastitis** in the herd or flock supplying the milk sample, and this warrants veterinary investigation.

Turbidimetric counting of microbes   Microorganisms have the ability to scatter light and when a critical number of microbes is reached, they cause sus-pensions to appear turbid. Within limits, the turbidity of a microbial suspen-sion is proportional to the number of cells present. The size and shape of cells, as well as their refractive properties, all influence the relationship between cell numbers and turbidity.

A rough indication of the number of microbes present in a suspension may be obtained by comparing the turbidity of the suspension with that of *McFarland's standard tubes*. These are prepared by mixing a 1% solution of anhy-drous barium chloride with a 1% solution of sulphuric acid. Depending upon the ratio of barium chloride to sulphuric acid used, different quantities of a fine, white precipitate of barium sulphate are formed. This causes each tube to have a standard turbidity, and this correlates approximately with the microbial count in a culture suspension (see Table 2.1).

There are two closely related methods for measuring light scattering more accurately. The **nephelometer** measures the amount of light scattered directly by using a photoelectric detector placed at right angles to the light source. **Spectrophotometers** measure the dispersal of light by placing the sample directly in line with both the light source and the photoelectric detec-tor. This measures the light lost from a culture after it has passed through the microbial suspension. The readings from a spectrophotometer are expressed as absorption or optical density units, measured at a specified wavelength. For microbial counting, a wavelength of 600 nanometres is often used. Suspension with an absorbance of greater than $0.5 \, A_{600}$ units are likely to need diluting with sterile suspension medium so that the linear relationship between cell numbers and optical density is maintained (Fig. 2.13).

Measures of light scattering to enumerate cell numbers can only be relative unless accompanied by a standard curve. Because microbes vary in light scat-tering properties, a standard curve must be prepared of turbidity against cell numbers. Standard curves must be derived using an alternative counting method, and using culture conditions identical to those of the current experi-ment. The standard curve calibration for one species under a particular set of conditions cannot be used to estimate cell densities of another species, nor of the same strain grown under different conditions.

Table 2.1 *Microbial counts using McFarland standard tubes*

McFarland standard number	Barium chloride	Sulphuric acid	Bacteria per millilitre
1	0.1	9.9	3×10^8
2	0.2	9.8	6×10^8
3	0.3	9.7	9×10^8
4	0.4	9.6	1.2×10^9
5	0.5	9.5	1.5×10^9
6	0.6	9.4	1.8×10^9
7	0.7	9.3	2.1×10^9
8	0.8	9.2	2.4×10^9
9	0.9	9.1	2.7×10^9
10	1.0	9.0	3.0×10^9

Wet and dry weight counting of bacteria   Microbial growth results in the production of novel **biomass**. Suspensions of microorganisms may be harvested by centrifugation, and the weight of the cells determined. A crude guide to the cell mass is made by measuring the *wet weight* of the deposit from the suspension. However, the quantity of liquid trapped in the pellet can vary and this may considerably influence the estimate of the cell mass. This problem may be resolved by determining the *dry weight*. To determine this value, the deposit is placed in a hot oven to drive off any water, and it is repeatedly weighed until a constant value is achieved. This represents the dry weight of the deposit, and is a better estimate of the cell mass. In a variation of this technique, a suspension of microbes is drawn through a pre-weighed filter, and the liquid is dried off under an infrared lamp. The filter is then reweighed, and the difference between the two weights is the dry weight of the microbes.

Microbial counts based upon chemical analysis   When microbes grow, they elaborate macromolecules from subunits present in the growth medium. The growth of microbial populations may be monitored by estimating the amount of protein, nucleic acid, or cell wall material present in a sample in comparison with sterile suspension medium.

Another chemical analysis that is used to estimate cell numbers is the measurement of ATP. This involves the use of the enzyme **luciferase**, which produces light from the hydrolysis of ATP. The more ATP present in a suspension, the more light is emitted. Bacterial cells all contain approximately

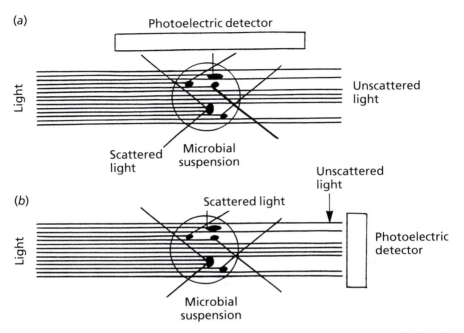

Fig. 2.13.   Scattering of light in (a) a nephelometer and (b) a spectrophotometer.

the same amount of ATP, and so the number of bacteria present in the sample may be estimated by measuring the **bioluminescence** of a suspension using luciferase.

Microbial counting using electrical impedance   When particles in suspension pass through an electrical field, they increase the electrical impedance. As microbes grow in a medium they alter the electrolyte balance in solution, and this, in turn, affects the electrical properties of the culture. The more microorganisms there are in a culture, the greater these effects will be. The microbial density of a culture can be estimated by placing widely spaced electrodes in the medium and measuring the electrical impedance. From this value, the number of microbes present may be obtained by reference to standard curves. This phenomenon is also exploited in the *Coulter counter*. Microbial suspensions are drawn through a tiny aperture, across which an electrical current passes. As microbial cells pass through the aperture, the impedance increases. The larger the cell or clump of cells passing through the aperture the greater the increase in impedance. With continuous monitoring, this permits not only the number of particles in suspension to be counted, but it also allows their size to be estimated. Thus, not only can an accurate total microbial count be rapidly determined, the degree of clustering of cells can also be estimated. This can be of value when correlating the total count to the number of colony forming units as determined by viable counting methods.

2.5.2 Viable microbial counts

A viable microbial count measures the concentration of living microorganisms present in a sample. In practice this may be difficult to define precisely. When a suspension of microbes is exposed to an adverse environment, it may take time for cells to return to full viability after a more hospitable environment has been restored. A widely accepted working definition of microbial viability is the ability to multiply and produce either a macroscopic colony on an appropriate solid growth medium or visible turbidity in an appropriate liquid culture. Other definitions may be applied in special circumstances. Exhibition of metabolic activity may be regarded as an indication of viability, even in the absence of cell division.

Viable counts using the spread-plate technique   The spread-plate technique is applied when the microbe to be counted grows best on the surface of the culture medium. It is also used if an opaque medium such as fresh or heated blood agar is required to support the growth of the organism. When using the spread-plate technique, the initial inoculum is subjected to a **serial dilution** to a point where no bacteria are expected to be present in the diluent. A tenfold, or decimal dilution series is most often employed. A small sample is then removed from each dilution tube and is placed on the surface of an appropriate solid growth medium. One Petri dish is used for each sample. The inoculum on each plate is then spread as a lawn over the entire surface of the plate using a sterile spreader. Often a sterile bent glass rod is used for this purpose. The cultures are then incubated to allow colonies to grow. The resultant colonies are counted and the number of bacteria present in the sample at the time the dilutions were made may then be calculated. The technique is error prone and takes a considerable time to perform, but it provides a sensitive estimate of the number of viable organisms in the culture at the time the sample was taken. By using a variety of selective media, the relative numbers of different organisms present in a microbial mixture may also be determined using the spread-plate technique.

Plates used for spread counts must be well dried so that the fluid from the inoculum is rapidly absorbed. In turn, this rapidly places cells in the inoculum in their final locations. This is particularly important when counting bacteria that are highly motile, and in the case of members of the genus *Proteus* even this may not be sufficient to prevent the bacteria swarming over the entire surface of many types of media. The swarming behaviour of *Proteus* sp. may only be stopped without affecting the viable counts by significantly increasing the concentration of agar in the medium, or by producing an electrolyte-deficient medium.

Many microbes are typically found in pairs, chains or clusters. A single colony from such organisms does not arise from a single cell. Rather, groups of cells giving rise to a single colony are referred to as **colony forming units**. It is often mistakenly assumed that colonies arise from single cells, and that by counting the individual colonies arising from a spread-plate, an accurate count of the number of microbial cells in the original sample is obtained. This is not the case. They do, however, indicate the number of colony forming units present. The use of a Coulter counter or direct microscopic examination may indicate the degree of clumping of a bacterium, and this may, in turn, reveal how accurate viable counting of the culture is likely to be.

After incubation, plates from dilutions of between 30 and 300 colonies are used to determine the number of colony forming units in the culture at the time of sampling. If fewer than 30 colonies are present, statistical variability renders the count unreliable. On plates with more than 300 colonies, competition for nutrients may lead to an underestimate of the viable count.

Viable counts using the pour-plate method   The pour-plate method involves the use of larger samples from each dilution than is used in the spread-plate method. The samples are mixed with molten, cooled batches of growth medium. Each dilution–growth medium mixture is then poured into its own sterile Petri dish. After the culture medium has set, plates are incubated as usual. Following incubation, the number of colonies both on and within the solid medium may be counted as for the spread-plate technique. This method is not suitable for heat-sensitive microbes, and since colonies form within the agar, obligate aerobes may exhibit a slight reduction in the apparent viable count when this method rather than the spread-plate technique is used. Another consequence of growth of colonies within the agar is that typical colonial morphologies such as are seen on surface cultures may not be apparent.

Viable counting using the method of Miles and Misra   A serial dilution of the suspension to be tested is prepared when using the method of Miles and Misra. The inoculum from each dilution is deposited as a drop onto the surface of a solid growth medium from a calibrated dropping pipette.

Each 20 microlitre drop is allowed to fall from a height of 2.5 centimetres onto the surface of the well-dried growth medium, where it spreads over 1.5 to 2 centimetres. Each of six plates receives a single drop of each dilution to be tested. Colony counts are made in the drop areas showing the largest number of discrete colonies that are not confluent following appropriate incubation. The mean count from the six plates gives the viable count per 20 microlitres of the dilution, and from this the viable count in the original sample may be calculated. This method was developed to overcome the cost of media employed in other techniques used for viable counting of microbes.

Viable counts determined using a spiral plater

A recent development in the determination of viable microbial counts is the *spiral plater*. This has the advantage that the method is semi-automatic, thus reducing operator errors. The spiral plater relies upon the principal of **dilution to extinction**; that is, diluting a sample until no more viable cells remain. A stylus delivers the inoculum onto the rotating surface of a solid growth medium. As the stylus moves out across the surface of the plate, the inoculum is being continuously and accurately diluted. After incubation, microbial growth is seen from the origin of the spiral. The greater the viable count of the original sample, the further along the spiral will growth be seen. Selective media can be used to count different organisms in a mixture.

Viable counts determined using filtration

When very small numbers of bacteria are expected in a sample that is relatively free from other particles, membrane filtration provides an excellent method of determining viable counts. Samples are drawn through a sterile filter whose pore size is too small to let bacteria through. The filter is then placed on the surface of a solid growth medium and incubated appropriately. The resultant colonies may then be counted.

Filtration gives good reproducibility of results and it allows samples of large volume to be examined. The technique is rapidly carried out, and may be performed in many field situations. It is also considerably cheaper than some of the other methods of determining viable counts, and has the advantage that selective media may be employed if this is desirable. Filtration cannot, however, be used for highly turbid samples because these block the filter easily. Furthermore, the method cannot be used for samples that contain heavy metal ions or phenolic compounds, since these are concentrated by being adsorbed by the filter material and thus inhibit microbial growth, lowering the estimate of the viable count obtained using this method.

Viable counting using the Most Probable Number technique   The Most Probable Number (MPN) technique combines dilution to extinction with statistical probability, based on the Poisson distribution, to estimate the viable count in the original sample. Dilution is achieved by taking decreasing volumes of the original sample as inocula for various tubes of liquid growth media. It requires a large original sample, is costly in media, and is technically more demanding than other counting techniques. By using a non-selective growth medium this method can provide information about the total microbial population present in a mixed sample, such as found in well-water. Substitution of a variety of selective growth media can yield information about groups of bacteria within the population. By increasing the stringency of growth conditions, counts of individual species may be obtained.

A 50 millilitre subsample of the original sample is used to dilute an equal volume of double-strength growth medium. Five 10 millilitre double-strength broths are each diluted with 10 millilitres of the original sample. Similarly, five 10 millilitre single-strength broths are inoculated with 1 millilitre of each of the samples, and five 10 millilitre single-strength broths are inoculated with five 0.1 millilitre subsamples of the original sample. All cultures are examined for growth following incubation under appropriate conditions. The probability of occurrence of microorganisms in samples follows the Poisson distribution, and hence the MPN of organisms present in the original sample is found by reference to statistical probability tables based on this distribution. When two different counts may be derived from the resultant pattern of growth, the 'best' result is that derived from the smaller sample volumes.

2.6   Growth of microbial populations

The ability to enumerate microorganisms permits microbiologists to study the growth of bacterial populations. Growth, however, involves separate but related processes. Microbial growth may be defined as an orderly increase in all components of an organism, normally associated with multiplication. This definition implies that at least two processes are occurring. Growth of a single microbial cell results in an orderly increase in its size. This process will continue until a critical point, when the second process, multiplication, occurs. The single, large microbial cell divides to yield two smaller daughter cells. In a few bacteria, and in many yeasts, cell division occurs by budding of the daughter cell away from the mother cell, but in the vast majority of bacteria cell division is achieved by a process known as **transverse binary fission**, i.e. splitting crosswise into two cells. Each daughter cell may then increase in size until the

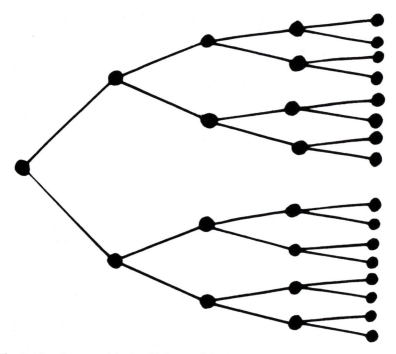

Fig. 2.14. Exponential microbial growth by binary fission.

critical point is reached once more, and cell division is again initiated, generating four cells. The next round of division will produce eight cells, each of which may grow and divide to yield sixteen cells, and so on (Fig. 2.14). Cell numbers in a growing population of bacteria increase in **geometric progression**. This is often referred to as **exponential growth**.

1 cell $\Rightarrow$ 2 cells $\Rightarrow$ 4 cells $\Rightarrow$ 8 cells $\Rightarrow \cdots \Rightarrow$ 1024 cells $\Rightarrow \cdots \Rightarrow$ 1 048 576 cells $\Rightarrow N$ cells

may be written as

2^0 cells $\Rightarrow 2^1$ cells $\Rightarrow 2^2$ cells $\Rightarrow 2^3$ cells $\Rightarrow \cdots \Rightarrow 2^{10}$ cells $\Rightarrow \cdots \Rightarrow 2^{20}$ cells $\Rightarrow 2^N$ cells $\Rightarrow$

The exponent, or power, of 2 increases by 1 at every cell division. In this way, as well as individual cells growing (i.e. increasing in size), the microbial population is also growing (i.e. increasing in numbers).

For simplicity, consider a culture derived from a single cell whose volume is

1 cubic micrometre. If cell division occurs every 20 minutes, and if growth was unrestrained, the volume of a culture of such a bacterium would be equivalent to that of an average sized house after only 22 hours and 40 minutes. After just 24 hours, the population will have undergone 72 divisions yielding approximately 10 000 000 000 000 000 000 000 cells, and after 48 hours, the population of bacteria arising from the single cell would have a mass about 4000 times that of the Earth. Common sense dictates that this does not occur, and that there must be constraints upon the growth of bacteria.

2.6.1   Batch culture of microbes

To understand the factors that control the growth of bacterial populations, the simplest experimental model is the batch culture. This comprises a liquid growth medium and a microbial inoculum that is permitted to grow with no net input into the culture except for the exchange of gases. There is no input of nutrients into the culture, and waste products are not removed. Under such conditions a batch of organisms is cultured.

At first, the bacteria or fungi present in the inoculum spend time adjusting to their new environment. During this phase, although individual cells may increase in size, there is no cell division. This period is known as the **lag phase**. Some physiologists regard the lag phase as extending only until cells increase in size, whereas others believe that it ends when the first cell division occurs. During the lag phase, microbes are undergoing considerable metabolic activity, and substantial phenotypic changes occur. The inducible enzymes required for active growth are synthesised during the lag phase, and the microbes become adapted to their new environment. The duration of the lag phase is highly variable, and depends upon the interaction of multiple factors. One of the most important of these factors is the previous history of the inoculum. If it comes from an environment similar to that of the new culture, then it will require little adaptation to its new environment, and this will not take very long. However, if it comes from a substantially different environment, the inoculum will require a considerable period of readjustment.

When microbes have adjusted to the new conditions in which they find themselves, they start to divide and the batch culture enters its **exponential** or **logarithmic phase**. Cell division is usually by binary fission, but some microorganisms do propagate themselves by budding. The doubling time and specific growth rate, described below, remain constant during this phase of growth. For most of the logarithmic phase, nutrients are present in excess of the demands of the microbial population. As the number of cells rapidly increases, eventually the nutrients in the growth medium become depleted, and toxic waste products accumulate. Apart from the exchange of gases, batch cultures are closed systems. This means that there is no input of nutrients and the products of metabolism are not removed from the culture. Eventually, a nutrient starts to run out, or a toxic waste product accumulates to the point where cells can no longer grow unrestrained. Growth slows up, and the specific growth rate diminishes to zero. When this occurs, the culture enters its **stationary phase**, during which the overall number of cells in the culture remains constant. The culture remains in a state of equilibrium during the stationary phase, and as cells die they are replaced by new cells resulting from controlled cell division.

It is during the stationary phase that so-called **secondary metabolites** are produced. These include pigments, **exotoxins**, and even antibiotics. It is also during the stationary phase that some bacteria produce spores in response to adverse environmental conditions. During the stationary phase, the rate of appearance of new cells is balanced by the rate of loss of cells from the culture.

Ultimately, the cells in a batch culture begin to die at a faster rate than they are replaced, and the culture enters the **death phase**. During this period, the loss of viable cells from a culture follows an exponential course. Autolytic enzymes are activated during this phase and the batch culture becomes progressively less turbid as dying cells lyse, until eventually the exhausted medium becomes clear. As with the lag phase, many factors interact to cause the transition from the stationary phase to the death phase. Both the stationary phase and the death phase may be of long or short duration, lasting from hours to weeks depending upon the microorganism being cultured, the type of medium in which it is grown and the conditions under which the culture is held.

Fig. 2.15 shows the four phases of growth represented as a graph of $\log_{10}$ cell numbers plotted against time. It can be seen from this graph that the transitions between each phase of growth are not clearly delineated. Rather, short periods of acceleration and deceleration of growth occur as conditions within the batch culture undergo rapid change.

In the model described above, during the logarithmic growth phase cell division occurs regularly if there are adequate nutrients and the culture conditions are kept constant. If this were to continue over a period of time, growth would proceed in a stepwise fashion, with the time taken between each cell division being represented as a horizontal line and the point of cell division being shown as a vertical line on the graph of $\log_{10}$ cell numbers plotted against time. This is shown in Fig. 2.16. This type of growth is referred to as

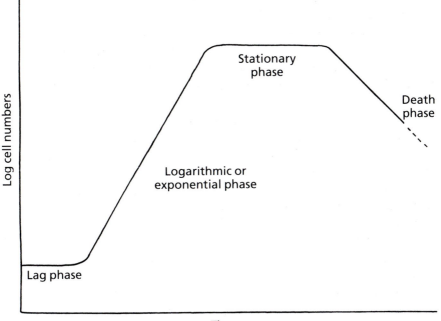

Fig. 2.15.   Phases of microbial growth in batch culture.

synchronous growth. In practice, synchronous growth requires special conditions, and can only be maintained for a limited number of cell divisions. Usually cell divisions are not precisely timed, so that the usual appearance of the growth curve is that of a smooth slope such as that shown in Fig. 2.15. The reason why $\log_{10}$ cell number is plotted against time is that if actual cell numbers were plotted instead of the $\log_{10}$ value, the graph would have a rapidly increasing gradient as seen in Fig. 2.17. Indeed, the time taken for the population to increase from 10 cells to 20 cells is the same as that taken to increase from 1 000 000 cells to 2 000 000 cells. This is a consequence of exponential growth. To convert this exponential curve into a straight line, a logarithmic scale is used for the vertical axis.

The **doubling time** (t_d) is defined as the time taken for the number of cells in a population to double. In addition to cells growing in numbers, microbial populations also increase in mass. Cell growth during the logarithmic phase is orderly and balanced. This implies that all the cells in the culture are in a similar physiological condition. Consequently, the doubling time refers to the time taken for the population to double in mass as well as in cell number. The growth

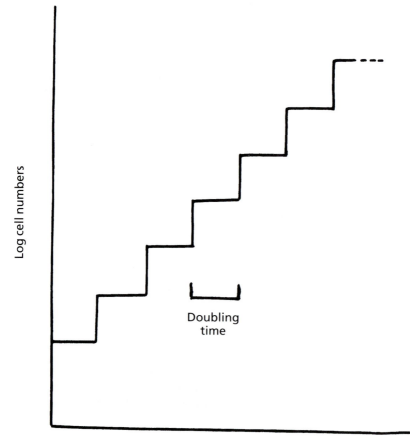

Fig. 2.16.   Synchronous microbial growth.

rate of a population is defined as the change in biomass with respect to time. The **specific growth rate** (μ) is a constant frequently used by microbiologists studying population growth. It is related to the doubling time by the formula:

$$\mu = \frac{0.69}{t_d}$$

From this formula it can be seen that as the specific growth rate increases, then the doubling time falls, because they are inversely proportional to one another.

Knowledge of the doubling time and specific growth rate allow predictions to be made about the behaviour of microbial cultures. In a rich growth medium, *Escherichia coli* may divide on average once every 20 minutes. Its spe-

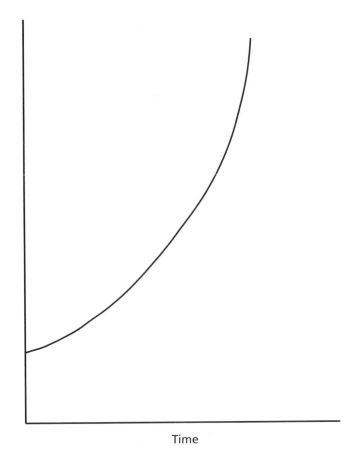

Fig. 2.17.   Exponential microbial growth plotted on an arithmetic scale.

cific growth rate $\mu = 2.07$ per hour (or hour^{-1}). If the same strain was growing in a nutritionally poorer medium, it may divide only once every hour, and so the specific growth rate $\mu = 0.69$ hour^{-1}. Even under ideal conditions, cultures of *Mycobacterium tuberculosis* double only once in every 18 hours, giving a specific growth rate $\mu = 0.038$ hour^{-1}. The specific growth rate can thus yield information not only about the organism but also about the conditions under which it has grown. Also, the smaller the value of μ, then the longer the culture will take to grow. This enables predictions about the growth behaviour of microbes to be made. Colonies of *Escherichia coli* should appear on a solid growth medium after overnight incubation, but because *Mycobacterium tuberculosis* has such a small specific growth rate it may take weeks for a colony of this bacterium to appear.

2.6.2   Simple continuous culture of microbes

For most purposes, batch cultures are perfectly adequate for cultivating microbes. However, within a batch culture, conditions are constantly changing as nutrients become depleted and as waste products accumulate. Consequently, microbes in batch culture are continually responding to their ever-changing environment. To study the physiology of a microorganism such a situation is undesirable. Preferably, external conditions should be kept as constant as possible so that the effects of manipulating a single factor can be observed without interference. To approach this ideal, **continuous cultures** have been developed. These maintain microbial populations in much more constant conditions than can be achieved in batch cultures. Furthermore the aeration, temperature, pH, etc. of the process can be continually monitored and adjusted as necessary.

To overcome the nutrient depletion of batch cultures, fresh growth medium is constantly added to a continuous culture chamber. Along with nutrient depletion, batch cultures undergo accumulation of waste products, and so in continuous cultures spent medium is removed at the same rate as the fresh medium is added. This has the advantage of removing not only waste products of metabolism, but also excess microbes. Consequently, the volume and biomass of the continuous culture vessel are kept constant and a **steady state growth** is maintained. This then permits single environmental factors to be altered, and their effects upon the physiology of microbes to be examined.

There are two principal types of continuous culture system in use, the **turbidostat** and the **chemostat**. In the turbidostat, the cell density of the culture is continually monitored photometrically, and whenever the turbidity of the culture reaches a critical point, the addition of fresh medium to the growth chamber is triggered. In the chemostat, fresh medium is continuously being added to the culture chamber and the rate of growth of the microorganism being cultured is regulated by the rate of flow of fresh medium into the vessel. Growth is regulated by the concentration of a limiting nutrient in this system (Fig. 2.18).

2.7   Artificial culture of fungi

Yeasts and moulds grow in artificial culture in a manner very similar to that of bacteria, and they display similar phases of growth. During the lag phase, the fungal cells adapt to their new environment and, after an unpredictable length of time, growth begins and then biomass increases in an exponential manner.

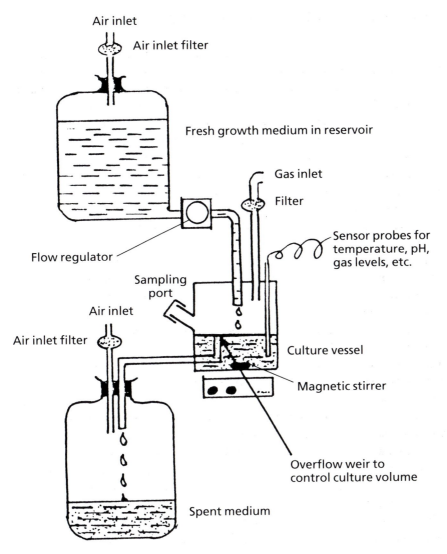

Air inlet

Air inlet filter

Fresh growth medium in reservoir

Gas inlet

Filter

Sensor probes for temperature, pH, gas levels, etc.

Flow regulator

Sampling port

Air inlet

Air inlet filter

Culture vessel

Magnetic stirrer

Overflow weir to control culture volume

Spent medium

Fig. 2.18. Continuous culture apparatus.

This represents the logarithmic or exponential phase of growth. The fastest rate at which hyphae double in length is about 2 hours. For yeasts, the fastest doubling time is about 20 minutes, but the typical doubling time for yeasts is nearer 90 minutes. This is slower than the doubling time of many bacteria. As nutrients run out or as toxic metabolic by-products accumulate, cell doubling slows, and the culture then enters a stationary phase. Again, there is a contrast with bacteria in that the cell density of bacterial cultures are often about 10^8–10^9 cells per millilitre. With yeasts, a much lower cell density is achieved; typically no more than 10^6 cells per millilitre. It is during the stationary phase of growth that secondary metabolites, including antibiotics, may be made. Eventually, the culture becomes exhausted and the fungus enters the exponential death phase.

On solid media, yeasts form discrete colonies similar to those formed by many bacteria. Yeasts are non-motile, and so in unshaken broth cultures, they form either a sediment at the bottom of the culture vessel or a film or **pellicle** on the surface of the growth medium. These different behaviours can be profoundly important. Beers and ales are fermented using brewing strains of *Saccharomyces cerevisiae*. These strains ferment vigorously at temperatures between 14 °C and 23 °C. The carbon dioxide that is produced carries the yeast to the top of the vessel, and hence these strains are referred to as top-fermenting yeasts. In contrast, lagers are produced by fermentation at lower temperatures, generally between 6° and 12 °C, and using a different sub-species of yeast, *Saccharomyces carlsbergensis*. This yeast forms aggregates that sediment, and is thus a bottom-fermenting yeast. The use of top- or bottom-fermentation is exploited to produce alcoholic drinks with very different characteristics.

In unshaken liquid cultures, moulds generally grow forming a surface mat, often with spores, but in shaken cultures they grow in three dimensions to produce a fungal ball. When growing on a solid medium, moulds grow predominantly radially and in two dimensions, although a degree of growth into the agar may occur. Aerial growth is of particular importance for fruiting structures bearing spores because moulds rely on the aerial dissemination of spores for dispersal.

Hyphae grow at the tip, or apex, of each filament, and this is termed *apical growth*. It involves a degree of cell wall lysis or breakdown to give plasticity in addition to the formation of new cell wall material. This process entails a balanced turnover of cell wall components. The growth process involves considerable cytoplasmic streaming towards the growing tip. The cytoplasm at the apex also becomes vesiculated. The vesicles are thought to deliver enzymes and cell wall components to the growing point, where the new wall becomes thickened, with the laying down of microfibrils. In filamentous organisms other than fungi, growth occurs throughout the length of the filament, and thus apical growth is a special feature associated with fungi.

Branching of the hyphae is necessary for the formation of mycelia. Hyphae form branches at points some distance from the growing apex. In order to form a new branch, a new growing apex forms within a mature portion of the hyphal cell wall. In septate fungi, branching is evident only from subapical cells. In two-dimensional growth on a solid medium, this leads to a structure reminiscent of a microscopic Christmas tree, as the branches grow away from the hyphae from which they were derived.

The life-cycle of a mould colony goes through distinct phases. Upon germination of a spore, vegetative mycelia are formed first. Later, asexual spores are formed and, as the nutrients become exhausted in the substrate medium, the colony will generate resistant, vegetative spores or sexual structures, if possible. Thus, sexual reproduction is in response to adverse external conditions, and is a method of re-assorting genetic material, increasing the probability of evolving structures that can survive such stresses when the latter are encountered at a later date.

2.8   Multiplication of viruses

Unlike the vast majority of bacteria and fungi that can grow on chemically defined media, the cultivation of viruses is complicated by the fact that they are obligate intracellular parasites. Their culture therefore requires the exploitation of living cells. These may be bacterial, or derived from fungi, plants or animals. At one time it was common to use whole animals or plants to cultivate viruses, but with the advances of tissue culture technology this practice is thankfully now largely redundant. However, the isolation and growth of viruses in tissue culture are expensive and take up a considerable proportion of the time that plant and animal virologists spend in the laboratory. Because of these limitations, many scientists who wanted to study viruses turned their attention to **bacteriophages**. The maintenance and growth of bacteria, whilst requiring a degree of technical competence and expertise, are faster and cheaper than experiments that use plant or animal tissue culture. Bacteriophages can be grown in bacteria maintained either in liquid culture or on a solid medium. Bacteriophage studies therefore dominated many of the early publications on the structure and replication of viruses.

In order to evolve and persist in their environment, all life-forms must reproduce. This is also true of viruses. Without reproduction, extinction would result. The reproduction of viruses is usually termed **replication**. This term has more in common with the biosynthesis of the macromolecules that aggregate together to form new virus particles, or **progeny**, than it has with the generation of new life-forms. Virus replication takes place in a susceptible host cell, and the series of events that take place from the entry of the virus into its new host to the subsequent release of progeny from the infected cell is termed the **virus growth cycle**. Virus growth cycles vary in length from less than 30 minutes for some bacteriophages, to several days for some large animal viruses.

There is no single pattern of replication for viruses, the events being different for different taxonomic groups. However, there are fundamental similarities between different virus growth cycles. All viruses must attach to specific receptors on new potential host cells. They must all pass their nucleic acid into the new host. The nucleic acid must be replicated, DNA genomes need to be transcribed, and the messenger RNA must be translated to produce new virus proteins. Finally, the new proteins must package the virus nucleic acid to assemble a new infectious particle. Viruses must also have the ability to interfere with the nucleic acid and protein synthesis of their host cell to allow the cellular efforts to be directed towards production of new viruses, rather than regular cellular components. The host cell typically dies as an outcome of virus infection. It generally bursts to release the newly formed **virions**. It is beyond the scope of this book to detail the many different strategies that viruses have adopted in order to replicate efficiently. However, it should be noted that the methods employed are very diverse.

It is possible to monitor the steady build up of virus within an infected cell by synchronously infecting a cell population, and examining them regularly at various times post-infection. This can be done in different ways, using electron microscopic studies, or biochemically by examining the appearance of virus nucleic acid or proteins, or by assaying for the presence of infectious virus particles. By plotting a graph of $\log_{10}$ virus particles per cell against time, a **one-step growth curve** of a virus can be generated. This is easily done in less than an hour when examining bacteriophage T4 growing in susceptible *Escherichia coli* cells. One-step growth curves are used to study bacteriophage growth, and a characteristic growth curve is shown in Fig. 2.19. For a time following infection, there is no increase in the number of bacteriophage particles, and this is referred to as the **eclipse period**. The **latent period** is the time taken from infection to the appearance of new virus in the medium. During the latent period, new bacteriophage particles are produced. As virus particles mature, they are released from lysed bacterial cells. This causes a dramatic increase in the number of bacteriophage particles in the culture, and this period is called the **rise period**. Once all of the infected bacteria have lysed, then the number of bacteriophage particles present in the culture reaches a plateau. The ratio of the number of bacteriophage particles at the beginning and the end of the growth curve gives the **burst size** of the bacteriophage.

For most viruses it is appropriate to divide the growth cycle into various stages or events. These include:

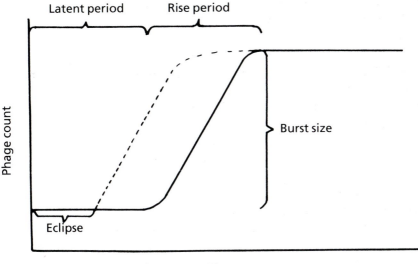

Fig. 2.19. One-step virus growth curve. The solid line shows the numbers of extracellular virus particles, whereas the dotted line represents the numbers of intracellular viruses.

1. Attachment of the virus to the host and penetration of the cell.
2. Biosynthesis of virus proteins and nucleic acids within the host cell.
3. Virus assembly and release.

The following discussion will be confined to a consideration of the replication of bacteriophages and animal viruses.

2.8.1   Replication of bacteriophages

Perhaps the best understood of the virus growth cycles are those of the T-even bacteriophages growing in cells of *Escherichia coli*. This family of bacteriophages (or phages for short), whose structure is shown in Fig. 2.20, contain double-stranded DNA in their heads. The length of the virus DNA is about 6% of the length of the *Escherichia coli* chromosome, and is sufficient to code for about 100 genes. Following a random collision between a bacteriophage particle and a bacterial cell, the virus is adsorbed onto the surface of the bacterium. The virus then attaches onto a specific receptor site on the bacterial cell via its tail fibre. Some bacteriophages attach specifically to pili or flagella,

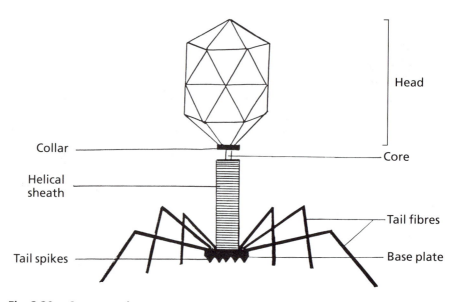

Head

Collar

Core

Helical
sheath

Tail fibres

Tail spikes

Base plate

Fig. 2.20.   Structure of a T-even bacteriophage.

and these are useful in studying the functions of such structures. The interaction between the bacteriophage and its host is chemical in nature, and chemical bonds are responsible for the attachment of the virus particle to the bacterial cell, rather than weaker forces. Following attachment, the bacteriophage tail releases an enzyme, *phage lysozyme*, and this breaks down a small area of the bacterial cell wall. The tail sheath of the bacteriophage then contracts, driving the tail core through the cell wall. As the tip of the tail core reaches the plasma membrane, the DNA from the head of the bacteriophage passes through the tail core and enters the bacterial cell (Fig. 2.21).

For most bacteriophages, the capsid structure remains outside the bacterium. This was exquisitely demonstrated by Hershey and Chase, who differentially labelled the protein and DNA components of bacteriophage particles. They found that only the label associated with the virus DNA could be found within the bacterial cell. The injection of virus nucleic acid into a host cell seen with the T-even bacteriophages is not common to all bacteriophages, and never occurs with animal viruses.

Once the bacteriophage DNA is in the bacterial cytoplasm, **biosynthesis** of virus nucleic acid and protein begins. The virus DNA takes over the metabolic machinery of the cell. Transcription of the host chromosome ceases, and its DNA becomes degraded. All of the messenger RNA that is tran-

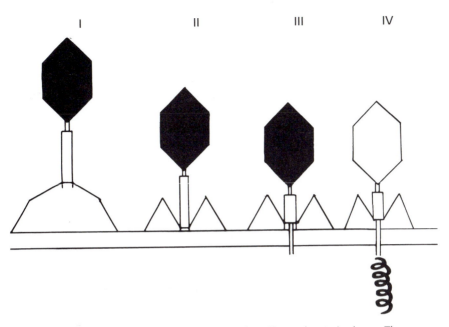

Fig. 2.21. Stages of infection of a bacterium by a T-even bacteriophage. The bacteriophage attaches to the surface of the bacterium via its tail fibres (I). The base plate is adsorbed onto the surface of the bacterial cell (II). The tail sheath then contracts, driving the tail core through into the bacterial cell (III), and DNA is injected from the head, through the tail core and into the bacterial cell (IV).

scribed at this stage is derived from the virus DNA template, and the bacterio-phage by now has totally taken over the bacterial cell for the purpose of its own biosynthesis and replication. However, previously produced host enzymes may continue to function and, for example, they still produce the energy necessary for the production of new bacteriophage nucleic acid and protein. Many bacteriophage-encoded enzymes are synthesised, and the two sets of enzymes work in conjunction to produce virus DNA, RNA and protein. The host bacterial cell provides the enzymes and amino acids that are used to produce the structural proteins of the nascent virus particles. Shortly after infection, however, no new bacteriophage particles can be detected within infected bacterial cells, since maturation of the newly synthesised particles can take several minutes. This short period within the virus growth cycle is referred to as the *eclipse period*, during which virus particles are not apparent.

As DNA and protein synthesis rapidly progress, large concentrations of newly formed virus DNA and structural proteins are produced. These then begin to assemble, or mature, into complete progeny virions. Therefore, within an infected cell there is a steady build-up of infectious virions that are readily detectable by breaking open the cell and using its contents in a plaque assay. The assembly process in bacteriophage T4 is highly efficient; much more so than in many viruses. The assembly of a new virion is guided by the products of bacteriophage genes in a step-by-step process. The bacteriophage head and tail are assembled separately from their respective subunits. The virus DNA is then packaged within the head and, finally, the tail of the virus is attached to its head.

Once a critical concentration of mature virions is reached within the infected cell, they are released into the environment by a process of cell lysis. In the case of bacteriophage T4 a virus-encoded enzyme, **lysozyme**, is synthesised within the bacterial cell, and this causes breakdown of the cell wall, leading to cell lysis. Like other virus proteins, the **transcription** of the messenger RNA responsible for lysozyme production is under **temporal control**. This means that it is produced at a specific time in the virus growth cycle. Because of the temporal control of messenger RNA synthesis, some proteins are produced early in the growth cycle, and others are produced at a later stage. The enzymes involved in virus DNA synthesis are produced from **early messenger RNAs**, and the capsid proteins and lysozyme that are produced later in the cycle are translated from **late messenger RNAs**. The time taken from the initial infection of a cell until its bursting is referred to as the **latent period**, and the number of bacteriophages released per cell is the **burst size** (Fig. 2.22).

The above cycle, resulting in the release of bacteriophage particles as a result of cell lysis, is referred to as a **lytic cell**. However, in addition to going through a lytic cycle, some bacteriophages may have their DNA incorporated into that of the host bacterium. They thus exist in a latent state without causing bacterial lysis. Such a state is referred to as lysogeny. The bacteriophages that are capable of undergoing lysogeny are called **lysogenic bacteriophages** or **temperate bacteriophages** and the host bacteria in which they reside are called lysogenic cells or **lysogens**. One of the most extensively studied lysogenic bacteriophages is bacteriophage λ found in *Escherichia coli* lysogens.

The DNA of bacteriophage λ when in the bacterial cytoplasm adopts a circular form that can go through a lytic cycle as described above, or it can recombine to integrate with the bacterial DNA in the cell. The inserted bacteriophage DNA is referred to as a **prophage**. Most of the prophage genes are silent. This means that they are not transcribed into messenger RNA that can be subsequently translated into virus proteins. The silence of prophage genes results from the production of bacteriophage-encoded **repressor proteins** that switch off most, but not all, of the virus genes. It would clearly be self-

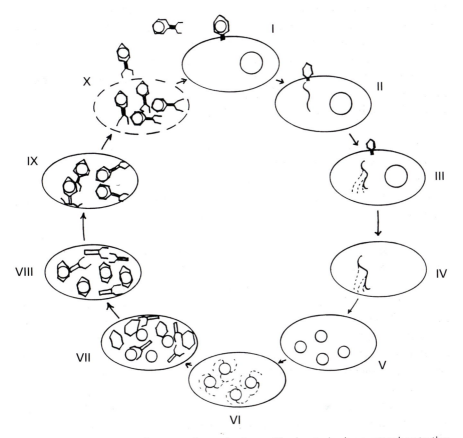

Fig. 2.22.   Replication of a T-even bacteriophage. The bacteriophage attaches to the surface of the bacterium to be infected (I). Bacteriophage DNA is then injected into the infected cell (II), and early bacteriophage messenger RNA is made (III). The bacterial chromosome is broken down (IV), and bacteriophage DNA is replicated (V). After this, late bacteriophage messenger RNA is transcribed and translated (VI), and components of the new bacteriophage particles are produced (VII). Bacteriophage DNA is then packed into the new bacteriophage heads (VIII). The new bacteriophage particles are fully assembled (IX), and are released by lysis of the infected bacterial cell (X).

defeating if they turned off their own synthesis! The prophage thus remains latent within a lysogen, and each time that the bacterial DNA is replicated for cell division, the integrated prophage is replicated as well. In this way, all progeny bacteria receive a copy of the prophage. Extracellular factors including particular chemicals, ultraviolet irradiation and temperature, or rare intracellular events acts as stimuli that result in the excision of the prophage from its host DNA. Once excised, the prophage may initiate a lytic cycle that results in the formation of many new virus particles.

Lysogenic bacteria are immune from infection with the bacteriophage that is present as a prophage, but they remain susceptible to other bacteriophages. Lysogenic cells often have particular properties bestowed upon them by their prophage. These result from the **transcription** and **translation** of certain prophage genes. The lysogens in which the prophage encodes a toxin are of great importance to humans. Strains of the bacterium *Corynebacterium diphtheriae* can cause diphtheria only if they are lysogenic, and if they carry a prophage that encodes diphtheria toxin. The strains that are not lysogenic are unable to cause diphtheria, since the toxin is required in order to cause the symptoms of the disease. Other important toxins that are encoded by prophages include the toxin of *Clostridium botulinum* that causes botulism, and the *Streptococcus pyogenes* toxin that is responsible for scarlet fever.

Whilst integration of virus nucleic acid into the host genome of an animal cell is a rare event, cases have been documented where such an event occurs as part of the virus growth cycle. The human immunodeficiency virus makes a DNA copy of its RNA genome, known as a **cDNA** copy, and this it inserts into the DNA of the host cell. This integrated DNA may remain latent for many years before it becomes activated. When this occurs, it initiates a virus replication cycle that ultimately leads to the clinical condition of AIDS. Other viruses may insert cancer genes, also called **oncogenes**, into the host chromosome, as with the Rous sarcoma virus that causes **sarcomas** in chickens. Alternatively, they may integrate next to oncogenes that are already present in the host genome and turn them on. This occurs with chicken leukaemia virus.

2.8.2   Replication of animal viruses

Animal viruses replicate by going through the same stages as those described above for the T-even bacteriophage. These include attachment and penetration, biosynthesis of virus nucleic acids and proteins, followed by virus assembly and release of the progeny. However, eukaryotic cells differ in fundamental respects from prokaryotic bacterial cells, and hence during their evolution animal viruses have had constraints enforced upon them by their host cell that do not apply to bacteriophages. Animal viruses have developed many and diverse strategies concerned with the organisation of their genes, their gene expression, replication of their genome and the assembly and release of new virions. These strategies have had to cope with the organisation of the eukaryotic cell with its plasma membrane, cytoplasm and membrane-bound nucleus, and with the partitioning of functions within the eukaryotic cell. DNA synthesis occurs only in the nucleus of a eukaryotic cell, and messenger RNA is transcribed only from DNA there as well. Protein synthesis occurs only in the cytoplasm. Furthermore, eukaryotic ribosomes can translate only **monocistronic messenger RNA**, that is messenger RNA that carries only one terminal recognition signal. In many cases, animal viruses have evolved to encode those functions that are not readily supplied for particular steps in the virus life-cycle because of the constraints of eukaryotic cell architecture. For example, poliomyelitis virus can synthesise messenger RNA from an RNA rather than a DNA template, vaccinia virus can cause DNA synthesis in the cytoplasm of a eukaryotic cell, and the human immunodeficiency virus can cause the synthesis of a DNA copy of its RNA genome. In a book of this size it is only possible to consider the generalities of animal virus life-cycles, and to give three examples of animal virus replication. These examples include a DNA virus, an RNA virus and a retrovirus.

Attachment of the virion to a susceptible cell involves interaction between a receptor and an anti-receptor site. One surface virion protein, or several proteins, known as **anti-receptors** will lock specifically onto one or more constituents of the host cell surface. These are the virus **receptors**. Receptors and anti-receptors are distributed throughout the surfaces of host cells and virus particles. Cell receptors are usually glycoprotein molecules. It is the cell receptor sites that establish the susceptibility of cells to virus infection. These determine the range of viruses that may infect the cells that support them. The specific interaction between a receptor and its anti-receptor can cause alterations in the configuration of the virion particle, and in some cases may lead to the rearrangement of specific molecules on the virion surface. These changes facilitate the penetration of a virion into its new host, and to uncoating of the capsid to release the virus nucleic acid. It has been proposed that this mechanism is involved in cell infection by human immunodeficiency virus. It is thought that the glycoprotein **gp 120** molecule on the virus surface acts as an anti-receptor that binds to the **CD4** receptors on its target cell surface. This stimulates the **gp 41** virus glycoprotein to induce virion fusion with the plasma membrane.

Penetration of the virus into its target cell occurs almost immediately following virus attachment. It appears to involve one of three mechanisms:

1. Translocation of the entire virus across the plasma membrane.
2. Receptor-mediated **endocytosis** of the virus particle, resulting in the accumulation of virus particles inside cytoplasmic vacuoles.
3. Fusion of the plasma membrane with the envelope of the virion.

Non-enveloped viruses such as the poliomyelitis virus penetrate cells when the entire virus translocates across the cell membrane. In this process, the capsid becomes modified in such a way as to allow cellular enzymes to degrade its structure, facilitating the release of the virus RNA genome for translation and the biosynthesis of new RNA strands. There is some debate concerning the precise nature of virus entry into an infected cell, and some scientists would include translocation across the plasma membrane together with **receptor-mediated endocytosis**. Some authors refer to this process as **pinocytosis**, and at present no clear consensus has been reached on this subject. **Enveloped viruses** such as influenza virus, herpesviruses and the human immunodeficiency virus penetrate cells by fusion of the envelope with the cell membrane. In such cases, the virus envelope remains as part of the plasma membrane of the target cell. The internal constituents of the virus are thus released into the host cell, where the nucleic acid is set free. The genome is released from the capsid by **capsid proteolysis**, and this is probably achieved using cellular enzymes rather than virus proteins. Some virus genomes remain attached to virion proteins during their transcription in the biosynthetic stage of the virus life-cycle.

Except for poxviruses, the nucleic acid of DNA viruses such as adeno-viruses, herpesviruses, and polyomaviruses is replicated in the nucleus of the host cell. Transcription also occurs within the nucleus, and the messenger RNA is transferred to the ribosomes of the cytoplasm for translation into virus proteins. These proteins may be transported back into the nucleus to assemble new virus capsids. Most RNA viruses, including the picornaviruses, paramyxoviruses and togaviruses replicate solely in the cytoplasm of their host cells. Synthesis of the virus RNA and translation of the virus messenger RNA takes place at different sites within the cytoplasm. Orthomyxoviruses such as influenza virus require nuclear factors early in their replication, although assembly of the progeny occurs within the cytoplasm. Retroviruses require their RNA to be converted into cDNA, and this is then transported into the cell nucleus, where it becomes integrated into the host cell DNA. Once certain viruses are assembled into capsids, in order to mature fully they acquire an envelope. This structure contains virus-encoded proteins and lipids derived from host cell membranes. Herpesviruses acquire their envelopes by the capsids budding through the nuclear membrane, whereas the human immunodeficiency virus obtains its envelope from the plasma membrane. In both cases, virus proteins are deposited in the appropriate membrane before the envelope is acquired by the mature virus. Capsid viruses such as poliomyelitis virus, leave their host cell by ruptures in the host cell membrane. This usually results in the death of the host cell. Examples of three virus replication cycles are presented diagramatically in Figs 2.23, 2.24 and 2.25.

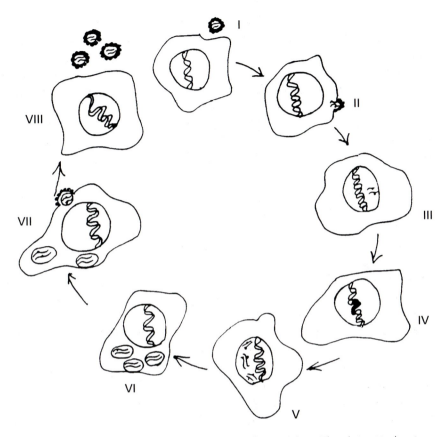

Fig. 2.23.   Replication of the human immunodeficiency virus. The virus attaches to the surface of the cell (I), and then enters the cytoplasm, where it releases its RNA (II). This moves to the nucleus (III). A DNA copy of the virus RNA genome is made using reverse transcriptase and this DNA copy becomes integrated into the cell genome within the nucleus of the cell (IV). The integrated cDNA copy of the virus genome is shown shaded. Eventually, virus RNA (V) and virus messenger RNA is made and transported to the cytoplasm where it directs synthesis of new virus particles (VI). The new virus particles are packed with a copy of the virus RNA genome (VII), and acquire an envelope by budding through the plasma membrane (VIII).

2.9   The effects of environment on microbial growth

Microorganisms are found world-wide. They can be isolated from the frozen wastes at the Poles, and have been found in the hot waters surrounding deep sea volcanoes. The minimum temperatures at which they can grow cannot be accurately determined, because of freezing problems encountered during the use of laboratory media. Bacteria from the vicinity of underwater volcanoes

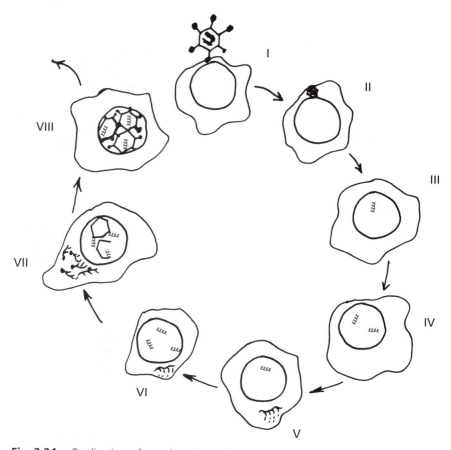

Fig. 2.24.   Replication of an adenovirus. The virion approaches the surface of the cell to be infected (I). It penetrates the cell, entering the cytoplasm (II). The DNA is released when the virus is uncoated, and the DNA moves into the nucleus of the cell (III). Early messenger RNA is formed (IV), and this leads to the production of non-structural proteins in the cytoplasm (V). Synthesis of virus DNA follows (VI), and then the virus structural proteins are produced and transported to the nucleus (VI), where the new virus particles are assembled (VII). Maturation of the nascent virus is followed by the release of the mature virus (VIII).

have been reported to grow rapidly at temperatures in excess of 250 °C and at a pressure of 265 atmospheres. The only requirement for growth would appear to be the availability of water in its liquid state. Microbes that have evolved to live at extreme temperatures can do so only because other environmental factors ensure that water is available as a liquid.

Because of their widespread distribution and their ability to exploit diverse and often hostile environments, it is perhaps surprising that microbial life-

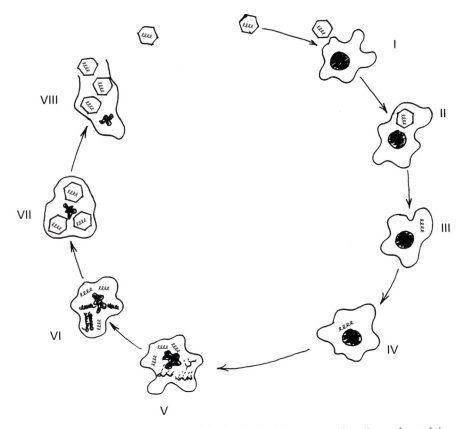

Fig. 2.25.   Replication of poliomyelitis virus. The virion approaches the surface of the cell to be infected (I). It penetrates the cell to infect it (II), and virus RNA passes into the cytoplasm (III). Non-structural proteins are formed during early translation (IV). The virus genome is replicated (V), and new virus structural proteins are formed (VI). New virus particles are assembled (VII), and released by lysis of the infected cell (VIII).

forms are not more obvious to the casual observer. They generally require microscopic observation of specimens, or cultivation in artificial media to become apparent. This raises the question of why this should be the case, and why is the world not knee deep in microbial cells. The reason for this is that in order for microbes to grow, like any organism, they require a supply of nutrients. In any environment, the supply of one nutrient will always be limiting. This is known as the *limiting nutrient*. If the supply of the limiting nutrient is increased, more microbes can be supported, and if it is decreased fewer microbes will be found. This principal can be seen in the artificial batch culture, but it applies equally to microbes in their natural habitat. If the supply of the limiting nutrient alters so that it no longer limits growth, then another substrate will become the limiting nutrient.

To cultivate microorganisms in the laboratory, they must be provided with adequate appropriate nutrients. They must also be placed in an appropriate environment to support growth. The conditions illustrated in the examples above provide evidence of the extremes of temperature that microbes may exploit, but other factors including the atmosphere under which a culture is grown, the pH of the growth medium and its osmotic properties all affect the ability of microbes to grow.

2.9.1   Temperature

All microbes have a characteristic temperature range over which growth is possible. There are three **cardinal temperatures** used to describe microbial growth. These are the **minimum growth temperature**, the **maximum growth temperature** and the **optimum growth temperature**. Microbial growth is maximal at an optimum temperature that reflects the temperature of the natural environment from which the organism is derived. **Psychrophiles** (literally meaning cold-lovers) grow well at 0 °C and have an optimum temperature of 15 °C or lower. The maximum temperature at which psychrophiles grow is about 20 °C. Some microbiologists refer to microbes that have an optimum temperature of between 15 °C and 20 °C with a maximum growth temperature of 35 °C as psychotrophs. Other microbiologists object to this terminology, meaning cold-eaters, on the basis that cold cannot be eaten! They prefer the term facultative psychrophiles.

Mesophiles grow in the temperature range of 15 °C to just above 45 °C. Their growth optima lie between 20 °C and 45 °C. Most human pathogens are mesophiles and have a growth optimum about 37 °C, normal human body temperature. Some human pathogens are killed if the body temperature is raised too far above normal. In the pre-antibiotic era, one treatment for syphilis was to induce fever in patients. One way in which this was achieved was to infect the patients with malaria parasites. This was considered an ethical treatment because **malaria** had a better prognosis than syphilis before the advent of effective chemotherapy. Not all human pathogens have a growth optimum of 37 °C. Campylobacters, the commonest bacterial cause of diarrhoae in the United Kingdom, have growth optima of about 45 °C. This reflects their natural habitat, the intestines of wild birds. Birds have a relatively high body temperature.

Thermophiles can grow at temperatures above 45 °C and often have an optimum temperature for growth in the range 55 °C to 65 °C. Some highly specialised thermophiles can grow at temperatures in excess of 100 °C, the boiling point of water. They can do so only at elevated pressures, so that the water in the environment remains in a liquid state. These bacteria cannot grow at normal atmospheric pressures and temperatures.

2.9.2   Atmosphere

Although atmospheric oxygen is essential for the life of many organisms, being used as an electron acceptor in cellular respiration, this is not necessarily true for all microbes. Microorganisms dependent upon the presence of oxygen are **obligate aerobes**. There are, however, microbes that are killed by the presence of atmospheric oxygen and these are **obligate anaerobes**. It should be noted that the spores of spore-forming obligate anaerobes are not killed by oxygen. Some bacteria are indifferent to the presence of oxygen, growing just as well in its presence as in its absence. These are known as **aerotolerant anaerobes**. The **facultative anaerobes** can grow aerobically or anaerobically, but their growth is enhanced by the presence of oxygen. The reason for this is that the facultative anaerobes can obtain energy from respiratory and fermentative metabolism. Cellular respiration, requiring oxygen, yields about ten times as much ATP as does fermentation; hence the growth of facultative anaerobes is enhanced by the presence of oxygen. There are a few microbes that are damaged by the normal atmospheric level of oxygen, but that require a small amount of oxygen (2–10%) to grow. These are the **microaerophiles**. Some aerobes, particularly intracellular parasites such as *Neisseria gonorrhoeae*, as well as needing oxygen require elevated levels of atmospheric carbon dioxide to grow. These bacteria are often grown in incubators flooded with 5–10% carbon dioxide, but the old method of cultivation was to grow cultures in candle jars. These are airtight containers in which a candle is allowed to burn until it extinguishes itself.

During oxygen metabolism, **superoxide radicals** are produced. The enzyme **superoxide dismutase** converts superoxides to hydrogen peroxide. This is broken down by the enzyme **catalase**. Obligate anaerobes lack both superoxide dismutase and catalase, and so fail to detoxify the oxygen radicals. Aerotolerant anaerobes produce superoxide dismutase, but not catalase.

Obligate anaerobes, although inhibited by oxygen, may frequently be isolated from sites that are apparently aerobic, for example in an **abscess** with a good blood supply. In such cases, they are frequently associated with facultative anaerobes that use up all the available oxygen, thus providing a suitably

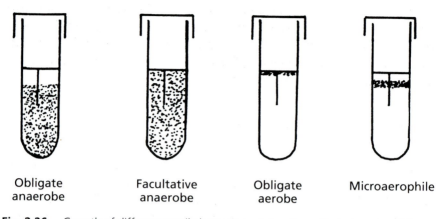

| Obligate anaerobe | Facultative anaerobe | Obligate aerobe | Microaerophile |

Fig. 2.26. Growth of different motile bacteria in stab cultures in sloppy agar. When an obligate anaerobe grows in sloppy agar, it cannot grow near the surface of the culture. Facultative anaerobes grow throughout the entire culture, whilst obligate aerobes can only grow at the top of the growth medium, where there is sufficient oxygen to support their culture. The growth of microaerophiles is confined to that region of the culture where the oxygen level suits their particular metabolic requirements.

reduced atmosphere to support the growth of the obligate anaerobes (Fig. 2.26).

2.9.3   pH

Most bacteria prefer to grow in a medium with an approximately neutral pH and these are referred to as **neutrophiles**. However, there do exist bacteria that grow in either extremely acidic or alkaline environments. These are the **acidophiles** and the **alkalinophiles**, respectively. The growth of some bacteria, such as *Vibrio cholerae*, in alkaline conditions can be exploited in selective culture techniques. Lactobacilli can grow at a lower pH than many bacteria, and the extreme acidophiles can grow in conditions analogous to placing one's finger in hot sulphuric acid. At the other extreme, some members of the genus *Bacillus* can grow in an environment of >pH 10. In contrast to bacteria, many fungi prefer a slightly acidic growth medium.

Whatever the preferences of microbes at the outset of culture, microorganisms have the ability to alter the pH of their local environment during the culture process. They do this by making either acid or alkaline products. The chemolithotroph *Thiobacillus thiooxidans* has a growth optimum of about pH 2, and produces sulphuric acid as a product of its metabolism. *Helicobacter pylori* is found in the stomachs of humans, a highly acidic environment. However, it can survive there because it elaborates a highly efficient **urease** that splits urea to yield ammonia. This raises the pH of the microenvironment, protecting the bacterial cells from acid attack.

2.9.4   Osmotic factors

Most microbes are hypertonic with respect to their environment. They have rigid cell walls and, being hypertonic, the cell membrane is pressed firmly against the cell wall. Because of the structure of their cell walls, Gram-positive bacteria can resist greater osmotic pressures than can Gram-negative bacteria. Bacteria achieve a hypertonic state by producing amino acids. Fungi produce glycerol, sucrose and mannitol for a similar purpose. If microbial cells are placed in a hypertonic medium they lose water, the plasma membrane shrinks away from the cell wall, and metabolic processes become disrupted. This process is known as **plasmolysis**. The ability of microbes to withstand osmotic pressures differs quite markedly. The osmotolerant nature of *Staphylococcus aureus* enhances its role in food poisoning. It can grow on salted foods in the absence of competing flora and, if it is an **enterotoxin** producer, it may go on to produce food poisoning.

Extreme halophiles, such as are found in the Dead Sea and in the Great Salt Lake in Utah, have become highly adapted in order to exploit a very hostile ecological niche. They have modified cell walls and membranes that are stabilised by the presence of sodium ions. They remain hypertonic to their environment by accumulating phenomenally high concentrations of potassium ions. If they are placed in media that is not salt-saturated, sodium ions are lost from the bacterial cell walls and membranes, which then simply disintegrate, causing cell lysis.

Potassium ions are accumulated inside cells as a first line of defence against adverse osmotic conditions. However, the continued presence of high intracellular potassium levels has a deleterious effect on microbes. To adapt to adverse **osmotic pressures** in the medium- and long-term, the cells manufacture **osmoprotectant** compounds such as betane. The regulation of production of such compounds is a fertile area of study, since it brings together the disciplines of microbial physiology and molecular biology.

3

Isolation, classification and identification of microbes

3.1 Isolation of bacteria

The earliest attempts to produce solid cultures included solidifying meat extracts with **gelatin**. Such media had two principal disadvantages: firstly, gelatin liquefies at about 37 °C, the optimum temperature for the growth of many human pathogens; secondly, many bacteria possess the ability to digest gelatin. Consequently, gelatin-based media tend to become liquefied under conditions where it is desirable to use a solid growth medium.

Agar is an inexpensive polysaccharide, obtained from certain seaweeds. In solution it can form a gel, and it provides an excellent substitute for gelatin as a solid support for microbiological media. Agar is generally resistant to microbial degradation and once gelled it remains solid at temperatures just below 100 °C. Once molten, agar suspensions remain liquid at temperatures of about 45 °C. This permits heat-labile supplements to be added to agar-based media without loss through heat degradation. Because of these properties, agar is the gelling agent used most widely in bacteriology.

During the earliest days of bacteriology, the most successful growth media were those derived from boiled extracts of meats of various types. Even now, brain–heart infusion broth is a rich growth medium often used to culture fastidious bacteria. Early culture media were very variable in their content. This caused problems with the standardisation of growth and also of bacterial characteristics. Today, many bacteriological growth media are still based upon **peptones**. These consist of a complex mixture of water-soluble products obtained from the hydrolysis of proteins derived from lean meats and other sources including heart muscle, **casein** and soya flour. Hydrolysis is generally achieved by digestion with proteolytic enzymes such as **papain, pepsin** or **trypsin**.

The main constituents of peptone include **proteoses**, amino acids, inorganic salts and vitamins. Because of the diversity of starting materials and the variety of modes of preparation, peptones vary considerably in composition. For some purposes, **tryptone** is used as a substitute for peptone. Tryptone is a tryptic hydrolysate of casein and is rich in tryptophan. Because the precise proportions of the components of these media are not known, they are termed *undefined media*.

Much effort has been directed towards ensuring consistency of undefined media and reducing batch-to-batch variation. Robertson's cooked meat medium originally contained minced beef heart, boiled in alkaline water and then suspended in a peptone broth. However, beef heart is variable in its constituents. Modern cooked meat broths employ textured vegetable protein spun in the form of minced meat, since it produces more reliable results. Consistency of growth media may be achieved by employing *defined media*. These are media in which the chemical composition is precisely known. Defined media tend not to support the luxuriant growth of bacteria as well as the complex, undefined media.

In their natural environment, bacteria grow in mixed populations with a diversity of species co-existing in dynamic equilibrium. In order to study the characteristics of one bacterial strain, it is essential to separate it from other microbes in the original sample. Failure to isolate the desired bacterial strain in the absence of contaminants leads to variable and unreliable results.

3.1.1   Separation of bacteria by mechanical means

Originally, all bacterial cultures were liquid-based, and the only way to isolate single strains was by a process of successive dilutions. This procedure was protracted and tedious. It yielded only the major components of the mixture in isolation, and the subculturing process was prone to permitting contamination of the culture. This was a particular problem when the growth medium employed was a body fluid such as urine, a very popular growth medium in the early days of bacteriology.

With the advent of solid growth media came the possibility of isolating bacteria with relative ease. The majority of bacteria grow as discrete colonies on the surface of solid culture media; these colonies are formed as a result of growth and division of a colony forming unit. It should be remembered that although a colony forming unit may comprise a single cell, it may also contain many cells in chains or clusters.

One of the most frequently used techniques for obtaining single bacterial colonies upon solid media is the *streak-plate* technique. Other methods include the *spread-plate* and the *pour-plate* methods. All three methods rely upon achieving spatial separation of colony forming units.

Single colonies are obtained using the streak-plate technique by the mechanical action of the inoculating loop as it picks up and deposits cells on the surface of a solid growth medium. By repeatedly dragging the loop across the plate, the original inoculum becomes diluted. The dilution effect is enhanced by the technique commonly employed in *plating-out*. An initial inoculum is made, and the loop is then sterilised. The loop is used to make a series of streaks from the initial inoculum. The loop is sterilised again, and a second series of streaks is made while the initial inoculum is carefully avoided. Once more the loop is sterilised, and another set of streaks is made, touching only the previous set. This process is repeated to produce the pattern seen in Fig. 3.1. In this manner, the original inoculum becomes sufficiently separated to give rise to single colonies. The spread-plate and the pour-plate techniques are described in Section 2.5.2 and both give rise to single, isolated colonies after appropriate dilution, inoculation and incubations have been carried out.

Both the spread-plate and the pour-plate techniques suffer from the fact that only the commonest bacteria in the original sample are isolated as single colonies. Minority species present in the original inoculum become overgrown. This problem can be partly overcome using streak-plates and repeatedly subculturing onto fresh media. This, however, is often unsatisfactory, and frequently results in failure. To overcome these problems, minority populations are often isolated using selective techniques.

3.1.2 Selective culture techniques

There are several ways in which to improve the chances of isolating a minority species from within a mixture of bacteria. This may be accomplished by treating the sample prior to plating-out, by manipulating the conditions under which bacteria are grown, or by the use of special media.

Spores are heat-resistant, and can resist boiling. To isolate a sporing organism, pasteurisation may be used. This involves holding the original mixture at a high temperature for a time sufficient to kill all the vegetative cells in a mixture, but leaving the spores undamaged. Holding mixtures at temperatures of between 65 °C and 70 °C for 30 minutes is usually effective.

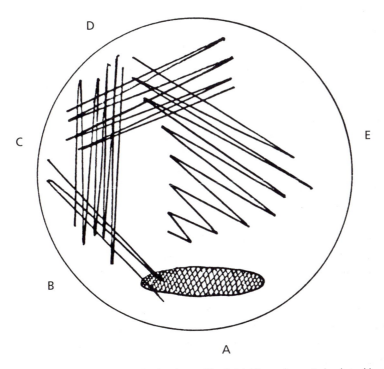

Fig. 3.1.   Plating-out of a microbial culture. The initial inoculum, A, is plated in a relatively small section of the plate. With a sterile loop, streaks are drawn out from this original inoculum to produce the pattern shown at B. The loop is sterilised, and a second pattern of streaks, C, is then produced, taking care not to re-enter the initial inoculum. This process is repeated to produce streak patterns D and E, again taking care only to streak from the immediately preceding streak pattern.

Some bacteria are remarkably resistant to extremes of pH, and this may be exploited in isolation procedures. Because of its waxy coat, *Mycobacterium tuberculosis* resists alkali. Sputum samples are heavily contaminated by bacteria from the upper respiratory tract and, if samples were not pre-treated with an alkali, these would rapidly outgrow the slow-growing mycobacteria. To overcome this problem, sputum samples are treated with 4% sodium hydroxide prior to culture inoculation. *Mycobacterium tuberculosis* is also acid-resistant. In the case of pulmonary **tuberculosis**, the causative agent has been obtained from stomach washings at pH 1. Viable bacteria are swept out of the lungs by the action of the **mucociliary escalator**, and are swallowed. Mycobacteria may then be removed from the very acidic stomach contents.

Legionella pneumophila is another pathogen of the lower respiratory tract. It is very fastidious in its culture requirements and like *Mycobacterium tuberculosis* it is easily outgrown by bacteria in the accompanying **commensal flora**. However, it is both heat- and acid-resistant. Specimen pre-treatment may involve either mixing samples with mineral acids or holding them at 50 °C for 30 minutes.

No incubation regime can be used to suit all bacteria. By manipulating the temperature and atmosphere under which cultures are grown, marked selective effects may be achieved. For example, psychrophiles must be grown at low temperatures and thermophiles require elevated incubation temperatures. They cannot be isolated in the same culture. Environmental organisms that are not derived from mammalian or avian sources tend to grow better at temperatures below 30 °C, whereas bacteria that have evolved to co-exist with warm-blooded animals grow better at slightly higher temperatures. Bacteria that live on or in humans tend to grow best at 37 °C, the normal temperature of the human body.

In addition to manipulation of the temperature of incubation, the atmosphere under which bacteria are grown is used for selective culture. Obligate anaerobes are killed in the presence of oxygen. Conversely, obligate aerobes require a continuous supply of oxygen. These two groups of bacteria, therefore, have mutually exclusive growth requirements. However, facultative organisms and aerotolerant anaerobes are capable of growth with or without oxygen, and so may be grown together with obligate aerobes or obligate anaerobes.

The pH of the growth medium may be adjusted to exert a selective effect upon a culture. Lactobacilli grow at a more acid pH than do most bacteria, and they may be grown on plates at pH 5. This inhibits the growth of many bacteria and is selective for lactobacilli. Similarly, *Vibrio cholerae* can grow in alkaline conditions. In faecal samples it may be outgrown by the vast numbers of other bacteria present. To enhance the chances of isolating *Vibrio cholerae* from a suspected **cholera** victim, stool samples are inoculated into alkaline peptone water at pH 8.4. *Vibrio cholerae* will happily grow under these conditions, but other bacteria in the faecal flora are unable to multiply. This dramatically increases the number of *Vibrio cholerae* in the culture, and these may be isolated by subculture onto an appropriate solid medium.

Manipulating the pH of a growth medium is one step towards producing a selective medium, and it should be remembered that by virtue of the pH and other factors all media are, to some extent, selective. It should also be borne in mind that microbes interact with their growth media to alter the pH as a result of their metabolism.

The selectivity of a growth medium may be enhanced by the incorporation of specific inhibitors. Solid media that incorporate inhibitors are simply referred to as **selective media**, but liquid-based media are frequently known as **enrichment broths**. This is because growth of the desired organism is not inhibited to the same extent as the growth of other organisms. Whilst other organisms are constrained by the inhibitor the desired organism will grow, and therefore its numbers will be enhanced within the mixed population. To isolate an organism from an enrichment broth, the culture is plated-out onto a solid growth medium after primary incubation. Often an enrichment broth is subcultured onto another selective medium.

The isolation of faecal pathogens provides an example of selective culture techniques. Often, faecal pathogens are greatly outnumbered by members of the commensal flora in the human bowel. Stool samples are plated on a primary selective agar such as MacConkey's agar. This medium contains bile salts, and these inhibit (some) non-enteric bacteria. If pathogens are present in sufficient numbers, they may be seen on the primary plates, but often this is not the case. Because of this, stool samples are also inoculated into selenite broth. This is a lactose peptone broth to which sodium selenite is added. Selenium salts are toxic to humans, animals and many bacteria, but salmonellae grow well in their presence. The number of salmonellae thus increases after growth in selenite, and they may be isolated by subculture of the selenite broth onto a selective agar-based medium. MacConkey's medium is often used in conjunction with a variety of other selective media.

As well as being a selective medium, MacConkey's agar is also an indicator medium. It contains lactose and neutral red, a pH indicator. Bacteria that ferment lactose produce acid, and the pH indicator turns an intense red. In turn, this colours the colony deep red. Bacteria that do not ferment lactose, referred to as non-lactose fermenters, consequently do not produce acid. Their colonies remain a buffish colour when grown on MacConkey's agar. Non-lactose fermenters can grow easily on MacConkey's medium, because it contains peptone as its principal nutrient. Most bacteria present in the aerobic commensal flora of the bowel ferment lactose, whereas faecal pathogens such as salmonellae and shigellae are non-lactose fermenters. *Shigella sonnei* can ferment lactose, but only after prolonged incubation. It is thus known as a **late lactose fermenter**, and after overnight culture appears as a non-lactose fermenter. Thus, colonies of potentially pathogenic species are easily spotted growing on MacConkey's agar. Further identification is necessary to differentiate salmonellae or shigellae from other bacteria unable to ferment lactose that may be found in the commensal bowel flora of some people.

Another selective medium that also acts as an indicator medium is mannitol salt agar, which is used in the isolation of staphylococci that are presumed to be pathogenic. A 5% solution of sodium chloride inhibits the growth of many bacteria, but not that of staphylococci or halophilic marine organisms. *Staphylococcus aureus* ferments mannitol to produce acid. This turns the phenol red indicator yellow. Other staphylococci are unable to ferment mannitol, and grow as reddish-coloured colonies. As with MacConkey's medium, the principal nutrient in mannitol salt agar is peptone.

Antibiotics and other antimicrobial agents may be incorporated into media to make them selective. *Neisseria gonorrhoeae*, the causative agent of **gonorrhoea** may infect the genital tract, the throat or the rectum. A rich and varied bacterial flora is normally present at each of these sites. *Neisseria gonorrhoeae* is also fastidious in its growth requirements and must be grown on a rich medium and under 5–10% carbon dioxide. In order to improve the chances of isolating *Neisseria gonorrhoeae* in the absence of competing flora, antimicrobial agents that inhibit many bacteria, but not neisseriae are added to the growth medium. Typically vancomycin, colistin, amphotericin B and trimethoprim are used in a rich medium incorporating lysed blood. Vancomycin inhibits Gram-positive bacteria, whereas colistin is active against many Gram-negative bacteria. Trimethoprim has a broad spectrum of activity and inhibits many bacteria not affected by vancomycin or colistin. Amphotericin B is used to inhibit fungi present in the specimen. *Neisseria gonorrhoeae* is resistant to each of these compounds, and is thus able to grow on VCAT agar, a selective medium containing all of these agents. VCAT agar is also used in the isolation of campylobacters from faeces. Campylobacters require a microaerophilic atmosphere for growth, showing that selective culture techniques may require the manipulation of multiple factors.

3.2   Isolation of fungi

The isolation of fungi in laboratory culture requires the growth of colonies on solid media, since in liquid cultures yeasts form either a sediment or a pellicle and moulds generally grow to form a fungal ball. In liquid cultures, sporing structures are apparent only if the fungus grows as a surface mat. Such structures are difficult to handle and it is hard to make preparations for microscopic examination from them. These problems are overcome by the use of solid media.

Cultures of fungi on solid media can be made either in Petri dishes or in tubes, where an agar slope is inoculated with the culture to be examined. Tube cultures take up far less room than Petri dish cultures and are less prone to aerial contamination with spores present in the laboratory air. However, because Petri dishes have a greater surface area, cultures are provided with more room for colony spread and development and the fungal growth is much more easily visualised and manipulated. Petri dish cultures also have the advantage of better aeration than tube cultures. This is an important consideration when cultivating obligate aerobes. However, some pathogenic moulds such as *Coccidioides immitis* or *Histoplasma capsulatum* produce numerous, easily displaced conidia in culture, and these fungi should always be grown in tube culture to minimise the risk of exposure and infection for laboratory staff.

Fungal cultures are, in general, maintained for much longer than bacterial cultures. Many bacteria give rise to colonies after overnight incubation, but fungal cultures can take up to three weeks to develop. During this time, loss of moisture from the solid medium in a Petri dish can be a problem. To overcome the difficulty of the medium drying out, an adequate quantity of agar medium must be poured into the plate. Typically, between 35 and 45 millilitres per 9 centimetre Petri dish is used for a fungal culture, whereas bacterial cultures rarely require more than 20 millilitres in a similar sized plate. A further measure to cut down water loss from Petri dish cultures is to seal the plate with an oxygen-permeable adhesive tape. The problem of aerial contamination is also reduced using this measure.

Great care must be taken when handling fungal cultures, particularly when moulds are being grown. Some pathogenic fungi must always be grown in tubes because of the very high risk of infection and the serious threat to the health of laboratory workers that they pose. Cultures should be opened only when absolutely necessary and be kept open for as short a time as possible. Not only does this help to reduce aerial contamination of that culture, it also helps to prevent the spread of spores from the colonies being examined. Observation of this precaution also protects the laboratory worker. Although many moulds may not be infectious, their spores can be highly **allergenic**. These can give rise to severe **allergic** reactions, especially if inhaled. You should never sniff a mould culture.

Many fungi grow at relatively low temperatures, with the optimum growth often occurring in the range 20–30 °C, although some fungi that are associated with warm-blooded animals including humans may also grow at slightly higher temperatures (≤37 °C). Although incubators are used to cultivate fungi, many cultures will grow quite happily, although perhaps more slowly, if maintained at room temperature.

Stock cultures are often maintained in screw-top McCartney bottles that contain agar slopes. Such bottles are also used for the transportation of fungi. It is very important that the caps are kept loose. Although the rubber stoppers in such bottles are permeable to oxygen, the atmosphere inside a bottle that has a tightly screwed cap soon becomes depleted of oxygen. This results either in the cessation of growth, or in abnormal growth and development of the culture. Cultures that are exposed to such stresses are frequently **pleomorphic**: that is, they display unusual structures or produce only vegetative growth and none of their characteristic spores.

3.2.1   Fungal culture techniques

In contrast to many bacteriological media, which have a neutral or slightly alkaline pH, fungi tend to grow best on an acid medium, typically pH 5.5–6.0. The pH of many fungal media selects against many bacteria, but the selective effect may be enhanced by the addition of antibacterial antibiotics such as chloramphenicol to the growth medium. Fungal growth media also tend to have a much higher carbohydrate content than do bacteriological media. This imposes osmotic stresses that again select against the growth of many bacteria. Because of the low pH of fungal media, agar tends to become hydrolysed at high temperatures. For this reason, sterilisation temperatures are not usually raised above 115 °C.

3.2.2   Media used in the cultivation of fungi

One of the most commonly used media for the cultivation of fungi in the medical laboratory is Sabouraud's medium. This is a peptone-based medium having a concentration of glucose that at 2–4% is significantly higher than those media used for bacteriological culture. Sometimes additional natural sources of complex carbohydrates are used to enhance growth or other characteristics of the culture. Complex media such as these also contain vitamins and other growth factors that enhance growth and proper development of the fungal culture.

Common examples of the other complex mycological media incorporating natural substrates include rice starch agar, used to stimulate the formation of mycelia and chlamydoconidia by *Candida albicans*, potato starch agar, malt extract agar and corn meal agar. Extracts of corn meal enhance the production of penicillin by members of the genus *Penicillium*. It was this discovery, made during the early 1940s, that enabled the first large-scale production of an antibiotic that has revolutionised clinical practice, even though the antibacterial properties of these moulds were described by Joseph Lister in the nineteenth century, 60 years before Alexander Fleming published his observations on the antibacterial effects of this fungus. When penicillin was first used on a patient, a policeman who was dying of overwhelming staphylococcal sepsis, the anti-biotic was in such short supply that it had to be re-extracted from the patient's urine. The benefits of using corn meal can thus be appreciated. During the period 1911–13, the laboratory attendant in the Botany Laboratory of the University of Cambridge was reported to collect unused cultures of *Penicillium glaucum* to use as a salve to treat skin infections. This folk remedy had been used in his family for a very long time. This was at least 16 years before Alexander Fleming described the activity of penicillin.

Some of the media used to cultivate fungi are synthetic and do not use complex organic components. Czapek Dox medium contains mineral salts with added sucrose. Like rice starch agar, this is also used to stimulate mycelial growth and the formation of chlamydoconidia by *Candida albicans*. This is enhanced by lowering the oxygen tension in the medium, and this can be achieved by placing a sterile coverslip over the edge of the inoculum. Some fungi grow less well on defined media such as Czapek Dox agar than on complex media, although Czapek Dox agar is better for the induction of spore production in many fungi.

3.2.3   Special techniques used for fungal culture

One of the principal aims of laboratory culture is to identify the fungi present in a sample, and a commonly employed taxonomic technique is to observe the spore structures that are produced by a mould. The preparation of a slide mount obviously causes considerable disruption to the structure of the fungus to be examined. This may cause severe difficulty in visualising all the necessary morphological features of the specimen. To overcome this diffi-culty, there are methods of visualising structures *in situ* without disturbing the fungal growth. The commonest of these are *slide culture* and *cellophane culture*.

Slide culture depends upon the hyphae and sporophores attaching them-selves to the glass of a coverslip or microscope slide, and thus the technique is not suitable for the examination of all fungi. To prepare a slide culture, a sterile glass slide is placed on a bent glass rod in a Petri dish. The glass rod raises the slide above a sterile, moist pad, used to keep the culture hydrated. A block about 1 centimetre square is cut from a solid growth medium that is known to support the growth of the fungus under examination. The block is then trans-ferred aseptically to the centre of the slide, and each edge of the square is inoculated with the fungus. A sterile coverslip is placed over the inoculated block, and the culture is then incubated. When sufficient growth has occurred, the coverslip is removed and a drop of alcohol is placed on the coverslip to dispel any air from the fungal material. The preparation is then stained with a drop of lactophenol cotton blue. The coverslip is then inverted over a clean slide and the culture is ready for examination under the microscope.

An advantage of the slide culture method is that two preparations can be made from one culture. The second preparation is made by removing the agar block from the supporting microscope slide, treating with alcohol, staining the preparation as before and covering the culture material with a clean coverslip.

Cellophane cultures exploit a similar process. Small pieces of sterile cellophane, about 1.5 centimetres square are placed on the surface of a solid growth medium, and are inoculated with the fungus to be studied. Fungi generally grow well on the cellophane support, obtaining their nutrients by diffusion through it. When sufficient growth has occurred, the cellophane square is removed and placed on a slide. Alcohol and then lactophenol cotton blue are added as above, and a clean coverslip is placed on top of the mount. It is best to use only a small inoculum when making cellophane preparations, since it is not possible to make preparations from large colonies grown in this way. Ideally a mixture of mycelia and spores is used to initiate the culture.

With certain aquatic fungi, agar cultures can be produced in the conventional manner, but spores are formed only in natural waters. To induce spore formation in these fungi, the culture has to be introduced into natural water. This is done using a 'bait', commonly a mixture of small seeds. Cress, onion or hemp seeds are first sterilised by autoclaving, and their seed coat removed with sterile forceps. These are added, together with the fungus to be studied, to a sterile mixture of coarsely filtered pond water and distilled water. The fungus will then grow on the bait, and may be easily retrieved from the culture for examination when the spore structures have had time to develop.

Fungi may be cultured from natural environments using baiting techniques. For example, fungi able to attack **keratin** substrates can be cultivated from soil by burying hair or even tennis balls in the earth. Similarly, leather, foodstuffs and other natural substrates can be used as bait to attract specific fungi. Indeed, it is not an uncommon practice to use leather and sterilised dung to cultivate fungi in the laboratory.

3.3   Laboratory culture and assay of viruses

Before detailed studies on viruses could be undertaken, a method for quantifying the number of virus particles in suspension was desirable. Indeed, for certain experiments, the need for quantitative results is essential. Plaque assays are a common method employed to assay viruses.

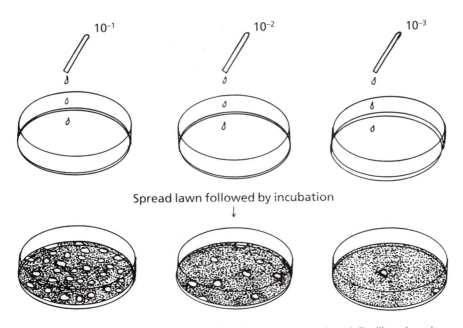

10^{-1} 10^{-2} 10^{-3}

Spread lawn followed by incubation
↓

Fig. 3.2.   Virus plaque assay. The bacteriophage suspension is serially diluted, and plated onto lawns of susceptible bacteria spread over the surface of Petri dishes. After an appropriate period of incubation at a suitable temperature the lawn of bacterial growth is examined for plaques; areas where there has been no apparent growth as a result of bacteriophage infection. If 0.1 millilitres of a 10^{-1} dilution yielded 200 plaques, 0.1 millilitres of a 10^{-2} dilution yielded 20 plaques and 0.1 millilitres of 10^{-3} dilution yielded 2 plaques, then the virus titre would be 2×10^{4} plaque forming units per millilitre. These figures are idealised: in practice the counts would show statistical variation.

3.3.1   Plaque assays

For bacteriophages and many animal viruses quantitative counts of virus particles may be achieved using a **plaque assay**. For a bacteriophage suspension this is a simple technique, outlined in Fig. 3.2. Briefly, the method is as follows. A diluted suspension of bacteriophage is mixed with a very dense suspension of bacteria in a fresh liquid culture. Typically 10^{8} bacterial cells per millilitre are used for a plaque assay. The mixture of bacteriophage and bacterial cells is combined with melted sloppy agar at 40 °C and poured into a Petri dish containing a solid layer of agar-based growth medium. The plate is then incubated to permit the bacterial cells to grow. During incubation the bacteria multiply to produce a 'lawn' of dense bacterial growth across the surface of the solid growth medium. During growth of the bacterial culture, the bacteriophages that were present in the original inoculum will infect bacterial cells in their proximity. After about an hour these bacteriophage-infected bacteria will lyse, or burst, releasing thousands of newly formed bacteriophage particles. The purpose of the sloppy agar in the **overlay medium** is to prevent the dispersal of virus particles. Instead, the progeny bacteriophages will infect adjacent bacterial cells. When this procedure is repeated as the lawn develops, small foci of infection will become apparent in the culture. Macroscopically, they appear as clear patches within the dense bacterial lawn. Such foci of infection are called **plaques**, and represent areas of the lawn where all, or most, of the bacterial cells have been lysed to release bacteriophages. Plaques can be easily counted. Each plaque is considered to have arisen from an infectious bacterio-phage particle, and therefore the concentration or **titre** of bacteriophage par-ticles that was present in the original suspension can be determined. Virus titres are usually expressed as plaque forming units per millilitre or p.f.u./ml for short.

Stocks of bacteriophage particles are usually grown by inoculating fresh liquid bacterial cultures with the bacteriophage particle to be cultivated, and incubating the culture overnight. The bacteria will grow in the liquid culture, but they will also become infected with bacteriophages. The viruses will repli-cate within the bacterial cells, causing them to lyse, and so, after overnight incubation, the culture will not look cloudy, as bacterial cultures appear, but rather will be clear as a result of the bacteriophage-induced lysis of the host bacterial cells. The clear overnight suspension of bacteriophage particles is centrifuged to remove any remaining bacteria, and the supernatant fluid, con-taining millions of bacteriophage particles, can be stored at 4 °C for several years with little loss of infectivity.

3.3.2   Animal culture

Animal viruses were at one time grown in animals such as mice, guinea pigs, rabbits and even primates. The use of whole animals is referred to as *in vivo* culture, from the Latin meaning 'in life'. These techniques were highly unsatis-factory and have now been superseded by the use of cell or tissue culture tech-niques. These allow living cells to be maintained outside the body in culture plates. Such techniques are referred to as *in vitro*, literally meaning 'in glass' even though plastic culture vessels are now commonly employed as well as glass. However, many laboratories do culture viruses in **embryonated eggs**. Influenza virus grows much better in embryonated eggs than in tissue culture. Therefore, to produce influenza vaccines, the virus is grown in eggs. If people

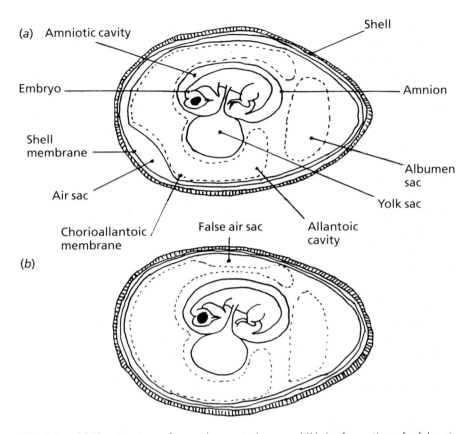

Fig. 3.3. (a) The structure of an embryonated egg and (b) the formation of a false air sac.

are **hypersensitive** to eggs, it is likely that they will show an allergic response when given the influenza vaccine.

An embryonated egg contains a series of membranes and fluids that are essential for the development of the **embryo**. These are illustrated in Fig. 3.3. The **chorioallantoic** membrane contains **epithelial cells**, and these provide an excellent support for the growth of the influenza virus. The influenza virus strain to be cultured is inoculated into the **allantoic cavity**, which is rich in fluid. The virus particles then infect the cells of the chorioallantoic epithelium, and many replication cycles produce thousands of new progeny virus particles. These are released into the allantoic fluid and can be easily harvested, since this fluid represents a virus suspension.

Embryonated eggs can also be used to assay for the infectivity of viruses. This is achieved by using a **pock assay**. Herpes simplex virus causes very evident pocks to form on embryonated eggs. Pocks are produced on the chorioallantoic membrane by creating a false air sac. This is achieved by sucking air from the actual air sac, causing the chorioallantoic membrane to drop. A dilute suspension of virus is then injected onto the membrane and the egg is then incubated. Virus particles infect cells within the chorioallantoic membrane, and these cells give rise to foci of infection. These foci then develop into pocks that appear macroscopically as round areas of **inflammation** on the membrane. Pocks contain many inflammatory cells in addition to cells that are infected. Counting the pocks on dissected chorioallantoic membranes gives an indication of the number of infectious virus particles in the original suspension, with the titre being expressed in pock forming units per millilitre (pk.f.u./ml). Care should be taken not to confuse this titre with plaque forming units or p.f.u.

3.3.3   Tissue cultures

Whilst being an essential part of the work carried out in particular virology laboratories, embryonated eggs are not widely used. Most virologists rely on the use of tissue cultures for animal virus cultivation. Indeed, it was the rapid development of tissue culture techniques that provided a spur to the rapid progress in virological research during the 1950s. An important example of work derived from the progress at this time was the development of vaccines against **poliomyelitis**. This could not have been achieved humanely or economically without the development of monkey kidney cell cultures, together with suitable growth media to support the growth of such cells outside the body.

Two vaccines have been developed that afford protection against poliomyelitis. The first to be produced commercially was the **Salk vaccine**. This employs dead virus particles to stimulate antibody production. The **Sabin vaccine** exploits live, **attenuated viruses** that infect and replicate in cells in the gut without causing disease. The Sabin vaccine was extremely popular when it was first produced, since it was administered on a lump of sugar. This was the first vaccination that did not require the use of a needle because sugar lumps provide an excellent way of delivering the live attenuated virus to its target, the gut. Following the introduction of these vaccines, the number of people suffering from paralytic poliomyelitis was dramatically reduced. However, even this success story is tinged with tragedy. The virus used in the Sabin vaccine carries mutations that render it incapable of causing clinical disease. The chances of these mutations reverting to wild-type, thus restoring the virulence of the virus particle, are several millions to one. Furthermore, in the vast majority of infections with poliomyelitis virus, the patient is either asymptomatic or suffers a mild gastrointestinal upset. Only in very rare cases does poliomyelitis virus cause paralytic disease. Unfortunately, there is a case recorded in which an unvaccinated person acquired paralytic poliomyelitis as a consequence of handling the soiled nappies of a baby who had been vaccinated with the Sabin virus, and in whom the mutations had reverted to wild-type. It is now medical practice not to vaccinate any child whose parents have not been vaccinated as well in order to avoid such a tragedy occurring again.

Animal tissue cultures are basically suspensions of animal cells grown in a culture medium in the laboratory. Cells grow on the glass or plastic surfaces of tissue culture vessels. They usually adhere to such surfaces by ionic bonding. Cell cultures are initiated at a low cell density. As cells divide on the surface of the tissue culture vessel **a cell monolayer** is formed. Some cell lines, including primary cell cultures and semi-continuous cell lines stop dividing once the monolayer is formed. Monolayers are formed because cells in culture demonstrate **contact inhibition**. Cells stop dividing once they are surrounded by other cells in the culture. This phenomenon is also seen *in vivo*. For instance, when scar tissue forms cell division stops once the wound is healed. In contrast to primary cells and semi-continuous cell culture, **continuous cell lines** do not exhibit contact inhibition, and they continue to divide once the monolayer has formed. This causes the cells to pile up upon one another. Continuous cell lines are often derived from malignant tumour cells that have also lost the phenomenon of contact inhibition. Again this loss is mirrored *in vivo*, since this helps to explain how tumours are formed by the unrestrained division of malignant cells.

Monolayers of cells will support the growth of animal viruses, given the constraints of virus tropisms. As described in Section 1.11, cell tropism is the phenomenon by which viruses can only infect a limited range of cell or tissue types. Growth of viruses in tissue cultures leads to biochemical and morphological changes in the cells within the culture. This usually results in the death of the host cell and the release of hundreds of new virions. The morphological changes that infected cells in culture undergo are referred to as **cytopathic effects**, or CPE, and these are easily observed with the aid of a light microscope. Many viruses cause specific changes in cells, leading to individual cytopathic effects. These can be used to identify viruses, and diagnostic virology laboratories frequently employ such observations in their protocols for virus diagnosis.

Tissue culture cells are essentially of three types. These are primary, semi-continuous and continuous cell lines. All cell lines are readily initiated by incubating dissected-out tissue with enzymes such as trypsin that cause the separation of tissues into individual cells that will form a new cell line. These cells are then enumerated, seeded out in a suitable growth medium and placed in a vessel suitable for cell growth. Commonly, the vessels used are either bottles or Petri dishes, and these are made of either glass or tissue culture grade plastic. Cell monolayers are formed by incubation of the culture at 37 °C and in an atmosphere in which the concentration of carbon dioxide is elevated to between 5% and 10%. Once a monolayer is formed, the cells are resuspended after enzyme treatment, usually with trypsin. Individual cells are released into the tissue culture medium, and these are used to re-seed fresh tissue culture medium that is then placed in a fresh culture vessel. The process of re-seeding tissue culture vessels is referred to as **cell passage** (pronounced *passarge*, as in French).

Primary cell lines are usually of mixed cell types, and frequently contain epithelial cells and **fibroblasts**. These are the first cells to be released from tissues such as human **amnion** or monkey kidney following **trypsinisation**. These tissues are widely used to initiate cell lines, since they can support the replication of a very wide range of animal viruses. They are of particular value in diagnostic virology laboratories, but unfortunately the cells die after only a few cell passages.

Semi-continuous cell lines are usually derived from aborted **foetal tissue**, since this has a greater capacity for maintaining cell division than cells isolated from adult animals. Human embryo lung, or HEL, cells, provide an example of cells that are commonly used to initiate semi-continuous cell lines. These cell lines are **cloned**, so that subsequent generations arise from a single cell. The cells remain diploid, retaining a full set of pairs of chromosomes. They can also be passaged between 60 and 100 times. This last characteristic makes them very useful in the virology laboratory. Unfortunately semi-continuous cell lines slowly lose their ability to become infected at high passage numbers, and are thus said to lose their sensitivity to infection. Semi-continuous cell lines have been of particular use in the culture of many human viruses such as the chicken pox virus, and also in the production of poliomyelitis virus vaccines.

Continuous cell lines are also referred to as **immortal lines**, and they appear to be capable of continued and indefinite cell passages. They have an abnormally high number of chromosomes that are not always in pairs. This condition is known as **heteroploidy**. They are also said to be **transformed**, since they have lost their contact inhibition. Most cells from continuous cell lines are capable of inducing tumour production if they are inoculated into susceptible hosts. The ease with which continuous cell lines can be passaged and maintained, coupled with their use to propagate a wide variety of viruses makes them a very popular choice for virus cultivation in most virology laboratories. Probably the most widely used continuous cell line was initiated from cells taken from a cervical carcinoma isolated from Helen Lane. She died from this cancer in 1951. This cell line, containing **HeLa** cells, has undergone thousands of passages. Its history and importance have provided the inspiration for television documentary programmes. Most virology laboratories will have HeLa cells, even if they are not routinely used but are only tucked away somewhere in deep freezers or under liquid nitrogen storage.

A very special type of tissue culture is referred to as an **explant**. Small blocks of tissue are removed aseptically from the organ to be examined, and then placed in an appropriate tissue culture medium. The culture is then incubated and cells grow out from the block. If these cells contain latent viruses the conditions in the culture trigger virus reactivation, and this can be observed as a cytopathic effect in the cells that emerge from the block.

Whilst the techniques for producing cell cultures were important for the advancement of animal virology, these advances could not have been made without parallel improvements in the development of suitable growth media in which cells could be propagated. In essence, growth media need to provide the correct isotonic conditions and nutrients at an optimum pH to support cell growth. Isotonic conditions are provided by a **basal salts medium**, and nutrients must include vitamins as well as glucose, amino acids and other nutrients. The correct pH in the medium is maintained by a buffer system. In order to maintain cells through numerous passages *in vitro*, the medium must be supplemented with serum. A variety of sera may be used, and commonly used sera include foetal calf serum, calf serum, rabbit serum or human serum. A fully supplemented medium is referred to as a *growth medium*. All of the above constituents have to be rendered sterile before use. This is achieved by membrane filtration. Contamination of animal cell lines by bacteria or fungi is a major thorn in the side of virologists and such contamination generally causes the death of the cells in culture. To help to overcome the problem of contamination many growth media are supplemented with antibiotics. Penicillin and streptomycin when used together provide a broad-spectrum antibacterial cover. Antifungal agents such as mycostatin are also commonly employed in tissue culture media. However, inclusion of antimicrobial agents in tissue culture media may adversely affect the cell line or virus propagation.

The size of vessels for tissue culture can vary depending upon the purpose for which the culture is to be used. When performing virus assays it is not uncommon to use a small polypropylene dish, measuring no more than 10 centimetres by 12 centimetres, and containing 96 small wells, each holding only about 300 microlitres of fluid. These plates can be found next to Winchester bottles that contain 2.5 litres and that are used to grow stocks of virus particles!

Animal viruses are assayed in a manner similar to that of bacteriophages, using the plaque assay technique. Diluted virus particles are added to a cell monolayer and are allowed to adsorb to the surface of cells for 30 minutes. After the adsorption period is complete the monolayer is covered with sloppy agar, or another suitable overlay medium. Derivatives of cellulose are commonly used for this procedure. Foci of infection develop within the culture, and these can be observed by the cytopathic effect that they produce in infected cells. Often, dead cells fall from the cell sheet, and an area of clearing appears within the cell monolayer. This has the appearance of a bacteriophage plaque. Such plaques are easily counted, hence the titre of the virus can be calculated by relating the size of the inoculum, and its dilution, to the number of plaques that are formed in a particular assay.

Virus stocks may be cultivated by inoculating a small number of particles onto a cell monolayer. Typically 0.01 plaque forming units per cell of the monolayer are used. After adsorbing the virus particles on the cells for half an hour, the cells are covered in growth medium and incubated. Usually after two or three days, the cytopathic effect develops in most cells of the monolayer. Typically, viruses will then have undergone two or three replication cycles. At this point virions may be present either in the growth medium or within infected cells. Virions are released from cells by smashing the cells open, then removing the cellular debris by centrifugation. The **supernatant** fluid contains the virus particles, and these can be assayed for infectivity and stored at temperatures between −20 °C and −70 °C. If animal viruses are stored at temperatures higher than −20 °C they quickly lose their infectivity.

Biological assays will estimate only the number of infectious units of virus in a particular suspension. However, during the replication cycle of many viruses, **defective virus particles** may be formed. These generally result from inefficient virus particle assembly. Defective virus particles may also arise as a result of thermal denaturation of previously active particles.

In order to determine the total number of virus particles in a suspension, electron microscopic techniques must be employed. Such studies are expensive and require considerable expertise. However, total virus particle counts are necessary if the quality of a virus suspension is to be determined. Following negative staining with phosphotungstic acid, virus particles can be readily visualised using an electron microscope. In order to count virions, **reference particles** must also be used. The most frequently used reference particles are latex beads. These are added at a known concentration to the virus suspension to be counted. The mixture is then prepared for electron microscopy, and the ratio of latex beads to virus particles is determined by visual observation of several randomly chosen fields. Since the concentration of latex beads in the suspension is known, then the total number of virus particles in the suspension may be estimated. For example, if there are twice as many beads as virus particles, then there are half the number of virus particles present in the suspension as there are latex beads. Of course, dilution factors must be taken into consideration when calculating the total number of virus particles in the original suspension.

Having estimated the total number of virus particles in this manner, this count can be compared with the number of infectious units in the virus suspension as measured by plaque assay or other biological techniques. The total number of virus particles compared with the number of infectious particles gives the **particle infectivity ratio**. This may be as low as 1:1 for bacteriophages such as bacteriophage T4. Alternatively, some animal viruses have a particle infectivity ratio as high as 1000:1. The non-infectious particles in these suspensions may have a role in disease. Defective particles may interfere with the replication of normal virus particles, and this may be important in the establishment of **persistent virus infections**.

Not all viruses can be propagated in tissue culture. This raises difficulties in identifying the infectious agents of particular diseases. Clinical conditions have been linked with non-cultivable viruses using electron microscopic studies. Other viruses may prove difficult to grow, and infections caused by these agents can be monitored serologically. Alternatively, viruses may be too dangerous to handle safely unless provided with the most stringent of containment conditions. All of these situations create difficulties for those studying virology. However, novel molecular biological techniques are being applied to help to overcome such difficulties. In particular, the **polymerase chain reaction**, or PCR, is being exploited to amplify specific sequences of DNA generated from viruses that are difficult to handle in other ways. Details of the polymerase chain reaction can be found in Section 3.5.4. To examine viruses that have an RNA genome, a cDNA copy of the virus RNA is made using the enzyme reverse transcriptase, and this cDNA copy serves as a template for PCR amplification.

3.4   Classification and identification of fungi

Classification and identification of organisms are two independent but interrelated processes. Classification is the division of organisms into hierarchical groups. These are based upon the degree of relatedness of organisms. Identification is the process whereby an isolate is assigned to a group within a classification scheme.

To illustrate the difficulties of classifying microorganisms, the parasite *Pneumocystis carinii* has, for many years, been regarded as a **protozoan**. It causes **pneumonia** in patients who are immunocompromised, and this is the most common life-threatening disease in patients with AIDS. It is a unicellular flagellate organism that lacks a cell wall. However, recent molecular genetic studies have shown that *Pneumocystis carinii* is more closely related to the fungi than to other protozoa. This is based upon the structure of its ribosomal RNA molecules, a feature that is commonly exploited in molecular taxonomy. Molecular genetic studies have therefore highlighted a situation where an organism cannot even be unambiguously assigned to its appropriate kingdom!

3.4.1   Classification of fungi

The classification of macro-fungi is well established, but in a book of this size it is not possible to include a description of the detailed classification and identification of macro-fungi such as mushrooms, toadstools, puff-balls or bracket fungi. Classification of such fungi is covered in many other books of a botanical nature. Rather, the text below concentrates on a description of the micro-fungi including moulds and yeasts.

Mycelial fungi, or moulds, are classified according to both their macro- and micro-morphology. Yeasts are structurally more simple and therefore display a limited range of morphologies. This creates difficulties in the classification of yeasts, and this group is subdivided partly on the basis of their reactions in biochemical tests.

There is no universally accepted classification scheme for fungi, but Table 3.1 outlines a commonly used scheme, modified from that originally proposed by G. C. Ainsworth in 1971. This permits the division of fungi into broad groupings and is the most widely accepted classification scheme available at present.

In some classification schemes the fungal divisions are referred to as phyla. Also, some have realigned the fungi grouped in the Phycomycetes. Members of the Myxomycota and Mastigomycotina have been regrouped into the kingdom Protoctista, which also includes protozoa and nucleated algae, and are not considered by some to be true fungi. Fungi in the Zygomycotina are retained as a group in the fungal kingdom and referred to as the Zygomycota.

The Myxomycotina or **slime moulds** are characterised by an amoeboid vegetative stage. However, under appropriate conditions the amoeboid cells congregate and differentiate to form reproductive structures that resemble other fungi. For this reason, members of the group including *Dictostelium discoideum* and *Physarum polycephalum* are intensively studied by developmental biologists. Slime moulds are common free-living organisms found in habitats such as leaf litter and soils, but some species are parasitic. The parasites are frequently found in association with higher plants, algae including marine algae, and other fungi. The symptomless parasite *Polymyxa graminis* is frequently found associated with the roots of cereal crops, and can act as the vector of virus diseases.

The Mastigomycotina are zoospore-forming fungi. They may form branched chains of cells that attach to their substrate by a root-like structure called a **rhizoid**. Many are soil **saprophytes** where they are found as important decomposers. Alternatively, they are found in freshwater habitats, and may be associated with water that is polluted with sewage. Some species are found as parasites of plants or algae, and a few are parasites of insects or fish. The downy mildews are obligate parasites unable to grow in standard laboratory cultures. The group includes important plant pathogens, such as *Phytophthora infestans*. This is the cause of potato blight. In the nineteenth century, Irish peasants used the potato as a primary food supply. When potato blight afflicted the potato crop in Ireland, the disease was responsible for the deaths of more than two million people during the Irish Potato Famine.

The Zygomycotina are common soil saprophytes and several species are associated with animal dung. Species of the genus *Entomophthora* are parasites of aphids and houseflies. The Zygomycotina also include a very important group of fungi that can form **symbiotic associations** with higher plants that are known as **mycorrhizas**. These structures involve the intimate association of a fungus and the root system of its associated plant. For example orchids have mycorrhizal associations in their roots. The fungus derives its organic nutrients from the plant and thus, in return, is provided with mineral nutrients that the mycorrhiza extracts from the surrounding soil. Fungi that form mycorrhizal associations may be impossible to grow in artificial culture.

The Ascomycotina include yeasts such as those of the genus *Saccharomyces*. These include *Saccharomyces cerevisiae*. This yeast forms the basis of the baking and brewing industries, and is of immense economic importance. Yeasts are frequently found associated with fruits, but can also be found in freshwater and marine environments. The mycelial Ascomycotina are common soil saprophytes, or are associated with animal dung. Fungi of the genus *Tuber* form mycorrhizal associations with the roots of trees. Their fruiting bodies are harvested as truffles that are highly prized culinary delicacies. In France,

Table 3.1 *Classification scheme for the micro-fungi*

	Divisions	Descriptions
Lower fungi (Phycomycetes)		
	Myxomycota (Slime moulds)	Organisms lacking cell walls, that nevertheless resemble other fungi in either lifestyle or behaviour
	Mastigomycotina	If formed, the mycelium is aseptate. This is also described as coenocytic. Asexual spores are formed endogenously in a sporangium. Spores and/or gametes are motile. Sexual spores show a varied form.
	Zygomycotina	The mycelium is aseptate. Asexual spores are found in sporangia. Spores are non-motile. Sexual reproduction occurs by gametangial copulation. Resting, zygospores are formed
Higher fungi		
	Ascomycotina (Ascomycota; Ascomycetes)	The mycelium is septate. Some members are unicellular (yeasts). Asexual spores, known as conidia, are borne exogenously. Sexual spores, called ascospores, are borne within a sac (ascus). Asci are borne singly or in groups within a fruiting body called an ascocarp
	Basidiomycotina (Basidiomycota; Basidiomycetes)	The mycelium is septate. Some members are unicellular (yeasts). If formed, the asexual spores are borne exogenously. Sexual spores, known as basidiospores are borne exogenously on a basidium, often within a fruiting body or basidiocarp. Basidiocarps are usually macroscopic structures

Table 3.1 (*cont.*)

Divisions	Descriptions
Deuteromycotina (Deuteromycota; Fungi Imperfecti)	The mycelium is septate. Some members are unicellular (yeasts). Asexual spores are borne exogenously, sometimes within a fruiting structure (pycnidium). Sexual reproduction is not known. Some coenocytic mycelial fungi with an unidentified sexual reproduction are included

pigs are specially trained to hunt out truffles by the smell that they emit. Not all of the Ascomycotina are benign. Dutch elm disease, responsible for the decimation of elm trees in England is caused by *Ceratocystis ulmi*, and the mildews of roses are caused by other ascomycete fungi. The **dermatophyte** fungi that cause diseases such as **ringworm** and athlete's foot are classified as members of the Ascomycotina if the sexual reproductive stage has been identified. However, many dermatophytes are classified as Deuteromycotina because their sexual reproductive cycle has not been observed and described.

The Basidiomycotina includes many fungi that live in association with plants. Some cause disease, but most are saprophytes that grow in leaf litter, composts, soil or dung. Fungi such as those of the genus *Agaricus* form fairy rings. Many of these fungi form mycorrhizas with trees. *Merulius lacrymans* is the cause of dry-rot in timber, and this single fungus is responsible for several million pounds sterling of economic loss each year. Basidiomycotina of the class Teliomycetes include fungi that are responsible for plant rusts or smuts, and these are also very important economically, since they frequently affect cereal crops. Those of the order Gasteromycetes, as the name implies, include the edible fungi.

The Deuteromycotina or Fungi Imperfecti, of necessity, include a wide variety of saprophytic and parasitic fungi. Many, like those of the genera *Aspergillus*, *Cladosporium* and *Penicillium* are important food **spoilage fungi**. *Aspergillus flavus* and related species of fungus are responsible for **aflatoxin** production. The presence of aflatoxin in foodstuffs is of great concern since aflatoxins are among the most powerful **carcinogens** so far discovered. *Aspergillus fumigatus* is responsible for the human disease aspergillosis, some forms of which cause serious and often fatal infection in immunocompromised individuals such as transplant patients. However, *Aspergillus niger* is of economic benefit, since it is used in the industrial production of citric acid. Similarly, members of the genus *Penicillium* are important in the production of antibiotics. *Penicillium chrysogenum* is used in the industrial production of the antibacterial penicillin family of antibiotics, and *Penicillium griseofulvum* is used to produce the antifungal agent griseofulvin.

3.4.2   Identification of moulds

The identification of a mycelial fungus involves the study of both macro- and micro-morphology. These will vary depending upon the growth medium and temperature used to cultivate the mould, hence it is often necessary to grow a fungus on a variety of media in order to complete its identification. The resulting colonies should be examined carefully, preferably with the aid of a hand-lens. Careful attention should be paid to the colour and texture of the colony. These features may be different in different areas of the colony, and the colour of a colony may be different on the surface and underneath the colony. The presence and nature of special structures including fruiting bodies, sclerotia etc. should be noted.

   Prior to disturbing the colonial growth to make microscopic preparations the entire culture may be examined using a low-power objective, so that fruiting bodies can be examined *in situ*. Lactophenol or lactophenol cotton blue mounts should then be made from the colony. A different mount should be made from each area of the colony that shows a different macro-morphology. Several mounts from each different area of the colony may be necessary to see all of the structural features associated with a particular fungus. Accurate observation is essential for the successful identification of a fungus growing in artificial culture. If a fungus has delicate sporing structures it is often necessary to use special culture techniques such as slide- or cellophane cultures that allow the micro-morphology of the fungus to be examined undisturbed.

   It may be difficult to differentiate fruiting structures such as pycnidia, perithecia, cleistothecia and sclerotia from each other. Some sclerotia are easily distinguished by their irregular shape. Others are not so easy to distinguish. The best way to identify these structures is to squash them to see what comes out, but sometimes sclerotia are difficult or impossible to squash. Sclerotia yield numerous oil droplets. Large numbers of conidia are released from squashed pycnidia, and asci and ascospores are released from cleistothecia and perithecia. Ascospores and asci tend to ooze out of their fruiting structures.

When all the observations necessary have been made, and the structures and their dimensions carefully noted, then the culture may be identified after reference to an identification key. There are several books that contain identification keys for fungi, but they are often devoted to particular groups. For example, there are books that are devoted only to penicillia or aspergilli. There is no all-embracing identification key for moulds. The following books contain keys that may be found useful:

Barron, G.L. *The Genera of Hyphomycetes from the Soil* (1968). Williams and Wilkins.

McGinnis, M.R. *Laboratory Handbook of Medical Mycology* (1980). Academic Press.

Samson, R.A., Hoekstra, E.S. & van Oorschot, C.A.N. *Introduction to Food-Borne Fungi*, 2nd edition (1984). Centraalbureau voor Schimmelcultures.

Smith, G. *An Introduction to Industrial Mycology*, 6th edition (1969). Edward Arnold.

3.4.3   Identification of yeasts

Yeasts are relatively simple structures and, paradoxically, this makes their identification more difficult than that of other fungi. Identification of yeasts is founded upon their limited morphological differences and on their biochemical properties. Morphological features used in the identification of yeasts include such things as cell size, shape and the presence or absence of a capsule. Some yeasts possess the ability to produce pseudohyphae, whilst a minority can produce a true septate mycelium identical with that found in moulds. Some yeasts are capable of sexual reproduction to produce ascospores. A small number of yeasts reproduce sexually to produce basidiospores. The capsulated yeast *Cryptococcus neoformans* that causes meningitis, particularly in patients with AIDS, is a yeast that belongs to the Basdiomycotina. *Candida albicans* is an important commensal and opportunist pathogen of humans. It can be identified from clinical specimens by its ability to produce a mycelial **germ-tube** when incubated for a period of 1–2 hours in serum at 37 °C. The biochemical tests involve observing the ability of the isolate to assimilate various compounds, mainly sugars and nitrogen sources, in addition to the ability to ferment different sugars. The identification of yeasts has been standardised and there are commercially available strips that can be inoculated to test for carbon and nitrogen assimilation. Depending upon the pattern of substrate usage an identification profile can be generated, and the identity of the yeast obtained either by reference to a manual or by using a computer database. The two most comprehensive guides to the identification of yeasts are Kreger van Rij's book *The Yeasts – A Taxonomic Study* and Barnett, Payne & Yarrow's book *The Yeasts*. Unfortunately their high cost means that they are reference volumes generally found only in specialist laboratories and libraries.

3.5   Classification and identification of bacteria

With higher organisms, the process of classification and identification are relatively easy, but bacteria are very simple structures and differ in very few ways. This makes bacterial classification a difficult task. However, with the rapid expansion of molecular biological techniques the classification and identification of bacteria is being considerably simplified. As a result, there has been a considerable revival of interest in bacterial taxonomy. These processes of classification and identification also have important applications in the everyday work of bacteriology laboratories.

It is important for bacteriologists to recognise the strains with which they are working and to differentiate them from organisms that may potentially contaminate their working cultures and stock collections. If this were not so the results of bacteriological experiments would be invalidated, since it would be impossible to tell whether the results were genuine, or an artefact resulting from contamination. Similarly, it is important that workers in different laboratories can be certain that they are working on the same strain when attempting to repeat a study. This requires the ability to identify bacteria.

Another example of the importance of strain differentiation is provided by hospital cross-infection incidents. Surgical wounds may become infected with *Staphylococcus aureus*. These infections are usually the result of human failure to observe good medical or nursing practice. Most often wound infections occur sporadically, but occasionally they occur in clusters. When clusters of infection happen, this is typically the result of one person's failure to observe proper practices. This problem is exacerbated by the fact that many hospital personnel carry *Staphylococcus aureus* as part of their commensal flora, and may shed large numbers of these bacteria into their environment. To investigate clusters of *Staphylococcus aureus* wound infections, isolates from each patient are collected, and all the staff on the unit are sampled for carriage of *Staphylococcus aureus*. All strains are then tested for their susceptibility to lysis by a collection of typing bacteriophages. The patterns of bacteriophage susceptibility are compared. Typically one member of staff will probably carry a strain that has the same susceptibility pattern as that found in the isolates causing wound infections in patients. In this manner the person responsible for the incident may be identified and his or her practices reviewed.

In most biological studies, the lowest unit of classification is the species. In the cross-infection example cited above, this degree of discrimination is inadequate because many people carry different strains of *Staphylococcus aureus*. To overcome this problem, differentiation at subspecies level is necessary. The same is true for many epidemiological and ecological studies.

3.5.1   Bacterial nomenclature

If bacteria are to be classified and identified, it is important that they are given a name. Each bacterium should have only one name. This name should be internationally recognised.

Among the first attempts to name bacteria systematically was assignation of the name of the discoverer to the isolate. Thus, when Robert Koch isolated the causative organism of tuberculosis it was called 'Koch's bacillus'. However, it quickly became apparent that this system of nomenclature was inadequate. What happened if a bacteriologist discovered *two* bacteria? There is still a tendency to honour eminent bacteriologists by giving bacteria Latinised forms of their names. Thus, for example, Bordet, Bruce, Erwin, Escherich, King, Lister, Neisser, Pasteur, Ricketts, Salmon and Shiga each have a bacterial genus named after them. Boyd, Stewart and Penner have had bacterial species named after them, and Morgan has had both a genus and a species named in his honour.

In the early days of microbiology, much interest centred upon bacteria that caused disease. Another system of nomenclature was to name the bacterium after the disease that it caused. Using this system, 'Koch's bacillus' became 'the tubercle bacillus'. Likewise, people who suffered **typhoid** were infected with 'the typhoid bacillus'. Recently, the causative agent of **legionnaire's disease** was placed in the genus *Legionella*. This system has also proved unsatisfactory because some bacteria cause several different disease processes. *Staphylococcus aureus* causes wound infections; it also causes toxic shock syndrome, osteomyelitis, brain abscesses, septicaemia, acute endocarditis, food poisoning, scaled skin syndrome, carbuncles, boils, impetigo and other skin conditions. Another problem with this system is that it ignores the majority of bacteria that do not cause disease.

Yet another attempt was to name the organism after its natural habitat. There is still evidence of this approach today. Thus, *Escherichia coli* is so called because it is found in the colon. Again this system falls down when trying to ascribe names to bacteria that are found in a diversity of locations.

In an attempt to resolve these problems, bacteriologists attempted to emulate botanists and zoologists by adopting a hierarchical classification system such as that proposed by Carl Linnaeus. Orders of bacteria are divided into families. In each family are a number of genera, each of which contains one or more species. Using this system, each bacterium may be ascribed a single binomial Latinised name, comprising the name of the genus to which it belongs, followed by its species name. The genus name is always capitalised, and after its first use in a text it may be abbreviated, often to just a single capital letter. By convention, the species name is written with a lower case initial letter.

Great care must be taken when abbreviating bacterial names. *Escherichia coli* is almost invariably shortened to *E. coli*, but where several common genera share the same initial letter then the first few letters of the genus name are used as the abbreviation. Thus, *Staphylococcus aureus* is frequently written as *Staph. aureus* and *Streptococcus pyogenes* becomes *Strep. pyogenes*. It should also be remembered that there is a genus of bacteria named *Bacillus*. The genus name, always written in italics and with a capital initial letter, should not be confused with bacillus, the description of any rod-shaped bacterium.

Bacterial names are printed in italic letters. In handwritten texts and documents that are produced on typewriters or word processors that do not have the facility to reproduce italic letters, bacterial names are usually underlined. This is the convention used by typesetters to indicate that text should appear in italics in its printed form. Italics are used only when one is referring to a specific organism such as *Staphylococcus aureus*. When talking about a group of bacteria, like the staphylococci, the name is not italicised, and if the group name is derived from a genus name then it is also not written with a capital initial letter. When discussing, in particular, the unnamed species within a genus, the name of that genus is given in full. It is written in italics, with a capital initial letter, and is followed by **sp.** or **spp.** depending upon whether one or several species are being discussed, e.g. *Staphylococcus* spp. that do not produce DNase are distinguished from *Staphylococcus aureus*. It should be noted that sp. or spp. are not italicised. Family names are Latinised and end in -aceae as in Enterobacteriaceae, and order names all end in -ales, for example, Spirochaetales. Latinised names may be of any gender. Thus, the genus *Staphylococcus* is male in gender, *Salmonella* is female and *Clostridium* is neuter. The plural forms of these genera are staphylococci, salmonellae and clostridia, following strict linguistic rules.

Currently, the International Committee on Systematic Nomenclature is the regulatory body controlling bacterial nomenclature. It publishes the

International Journal of Systematic Bacteriology. In 1980 it published a list of approved bacterial names. Any additions to this list, or alterations of previously adopted names are published in this journal after submission to and approval by the International Committee.

As an inheritance from the older systems of nomenclature bacterial names are often descriptive, reflecting a key feature of the bacterium. Names may indicate a disease process with which the bacterium is associated. *Salmonella typhi* causes typhoid in humans, *Mycobacterium tuberculosis* causes tuberculosis, *Shigella dysenteriae* causes bacilliary **dysentry**, and the recently described *Legionella pneumophila* causes legionnaire's disease. Other names reflect the colonial or microscopic appearance of the bacterium. *Staphylococcus aureus* is the golden (*aureus*) berry (*coccus*) that clusters like grapes (*staphule*), *Micrococcus luteus* is the small (*micro*) yellow (*luteus*) berry (*coccus*). Names, however, can be misleading. For a long time *Haemophilus influenzae* was thought to cause influenza. It is now known that influenza is caused by a virus infection. *Haemophilus influenzae* is isolated from many patients recovering from the disease because it causes secondary infection of tissues already damaged by the influenza virus. Perhaps the most misleading species name of all belongs not to a bacterium, but rather to humans: *Homo sapiens*, the wise man!

3.5.2   Bacterial taxonomy

Bacterial taxonomy has recently entered a phase of rapid and radical change as the new techniques of molecular biology are being applied to this field. Taxonomists are constantly refining the systems used for classification of bacteria. There has also been much more care taken recently in ascertaining the history of a species so that species names reflect the original description, rather than a popularly used name. As a result of such activity, the species names assigned to bacteria may change.

For many years the bacterium causing gas gangrene was called *Clostridium welchii*, but it was decided that the first description of this bacterium was not attributable to Welch, but, rather, it was first described as *Bacillus perfringens*, the perforating rod. This description dated from the latter part of the nineteenth century, at a time when all rod-shaped bacteria were named *Bacillus*, and long before the genus *Bacillus* as it is now understood was proposed. Because its true pedigree has been established, the name of this bacterium has now been changed to *Clostridium perfringens*. Name changes such as this may cause problems when older literature is being reviewed.

Bacteria may suffer a series of species name changes. There are bacteria that are associated with stomach ulcers. These were first described by histologists studying gastric biopsies. This was before bacteriologists had managed to cultivate them. On the basis of their morphology they were called campylobacter-like organisms. When bacteriologists did manage to cultivate these bacteria they did, indeed, behave like campylobacters. Because they came from a region around the pylorus the species was named *Campylobacter pyloridis* (Greek: *campylo*, curved; *bakterion*, rod; *pyloros*, gate-keeper or warden). Therefore *Campylobacter pyloridis* is the curved rod of the gate-keeper. Although *pyloros* is a word of Greek origin, it was adopted into Latin as *pylorus* and the correct genitive case of the noun is *pylori*. The International Committee on Systematic Nomenclature thus made a linguistic error in accepting the name *Campylobacter pyloridis*, and it was renamed *Campylobacter pylori* in order for the name to be linguistically correct. Shortly after this name change, molecular studies on the structure of the 16 S rRNA gene showed that this bacterium was not very closely related to other bacteria then in the genus *Campylobacter*. Indeed, the differences between *Campylobacter pylori* and the other campylobacters were sufficient to warrant placing it in a separate genus. It therefore became the type species of the genus *Helicobacter*, and is now called *Helicobacter pylori*. The time taken for campylobacter-like organisms to become *Helicobacter pylori* took less than 5 years.

With higher organisms, there is little question about what constitutes a species. It is a group that shares many common features and its members breed to produce fertile offspring. This concept is meaningless in bacteriology, where bacteria reproduce asexually by binary fission. Furthermore, the genetic exchange mechanisms that do exist in bacteria permit DNA to be mobilised through a vast array of unrelated bacteria. Moreover, bacteria have very few characteristics useful for comparison, and thus the division of bacteria into species is an entirely arbitrary process.

A collection of similar species constitutes a genus. Just as there is difficulty in deciding what constitutes a bacterial species, so there is no satisfactory solution to the problem of what constitutes a bacterial genus. Until recently, bacterial taxonomy has been a somewhat arbitrary process. Now these matters are strictly regulated by the International Committee on Systematic Nomenclature.

The introduction of the use of computers in bacteriology has had a fundamental impact on bacterial taxonomy. It has permitted the widespread adoption of the techniques of numerical taxonomy. This is alternatively named Adansonian taxonomy, after the eighteenth century French botanist who pioneered the system. This approach involves comparing strains of bacteria for the expression of a large number of different characteristics. The more closely related organisms are, the more characteristics they will have in common. A **similarity matrix** can be drawn up showing the degree of relationship between all the species examined. Having constructed a similarity matrix, the results are often presented in the form of a **dendrogram**. The degree of relatedness is represented along the x-axis, and nodes occur at the points of similarity. A dendrogram is illustrated in Fig. 3.4.

Numerical taxonomy compares the **phenotypes** of organisms, and is sometimes described as a **phenetic** system. It relies upon the expression of particular genes rather than just their presence. It represents a close approach to the description of the evolutionary relationship of organisms, but it is not completely satisfactory because it cannot account for the simple presence of genes.

Since Charles Darwin developed the theory of evolution by natural selection, biologists have tried to devise **phyletic** systems of classification in which the true evolutionary relationship of organisms is represented. In an attempt to determine the phyletic relationships between bacteria, taxonomists are turning their attention to a direct analysis of their genetic material.

Early studies were very crude and yielded little information about evolutionary relationships. At first, the melting temperature of DNA isolated from different bacteria was determined. From these values, the relative proportions of guanine plus cytosine base-pairs, or the G + C content, for each strain may be calculated. The more disparate the G + C contents of two bacterial species are, the more distantly they are related. Caution must be applied to the results from such experiments. Just because two strains have similar G + C contents, it does not mean they are closely related. It may be that one evolved from an ancestor with a high G + C content, and the other evolved from an ancestor with a low G + C content. Over the course of time their evolution may have converged, so that they now have similar G + C contents, even though this was not always the case.

An advance on melting temperature studies was the development of DNA hybridisation studies. Originally the degree of hybridisation was measured by determining the temperature at which a mixture of melted DNA strands isolated from different bacteria **re-annealed**. This was performed by observing changes in the ultraviolet absorbance of the DNA mixture at various temperatures.

DNA hybridisation techniques have been refined, and **Southern blotting** and hybridisation techniques are now often used to indicate the degree of relatedness of different DNA sequences. Southern blotting was a technique developed by E. M. Southern, hence its name. DNA fragments from one bacterium are separated electrophoretically on an agarose gel, and denatured to

	Organism				
Test	A	B	C	D	E
1	+	+	−	−	−
2	+	−	+	−	−
3	+	−	−	−	−
4	−	−	+	−	+
5	+	+	+	+	+
6	−	−	+	+	+
7	+	+	−	−	+
8	+	+	−	+	+
9	−	−	+	−	+
10	−	−	−	+	+

Test results are compared pairwise

	A	B	C	D	E
A	1.0				
B	0.8	1.0			
C	0.3	0.3	1.0		
D	0.4	0.6	0.5	1.0	
E	0.3	0.5	0.6	0.7	1.0

From this comparison a similarity matrix is constructed

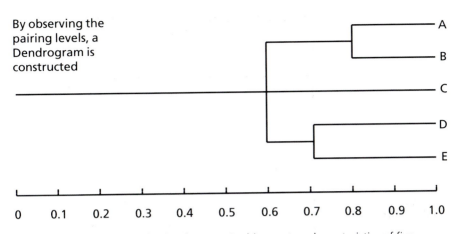

By observing the pairing levels, a Dendrogram is constructed

Fig. 3.4. Construction of a dendrogram. In this case ten characteristics of five individual organisms are compared pairwise. Organism A resembles organism B in eight of the ten characters being observed. These include both positive and negative results. From this it can be seen that organisms A and B have a similarity of 0.8. Likewise, organism A and organism C share only three characteristics, and thus they have a similarity of 0.3. From such comparisons, a complete similarity matrix may be constructed. At a similarity of 0.8, organism A pairs with organism B. At a similarity of 0.7, organism D pairs with organism E. At a similarity of 0.6, organism B pairs with organism D and organism C pairs with organism E, thus at this level every organism tested has paired with at least one other organism examined. By identifying the pairing levels, a dendrogram may be constructed.

produce single-stranded DNA molecules, rather than the familiar double-stranded form. These are then transferred to a filter, usually made of nitrocellulose or nylon. After attaching the single-stranded DNA to the filter by baking, denatured DNA from a second bacterium is labelled either radioactively or chemically, and it is allowed to react with the filter-bound DNA. If the two bacterial strains have DNA in common, then the labelled DNA will form a **heteroduplex** with the filter-bound DNA. This is a double-stranded structure in which one strand is derived from each bacterium. Unbound, unlabelled DNA is removed by washing the filter thoroughly and the bound heteroduplex can be detected, either by **autoradiography** or by chemical detection, depending upon the method of labelling.

The labelled DNA probes used in Southern hybridisation do not need to be an entire bacterial genome, and may only be a small sequence from within a gene. Indeed, oligonucleotide probes may be as short as 10–15 bases. Oligonucleotide probes tend to be used more for identification than classification purposes (Fig. 3.5).

One drawback with DNA hybridisation studies is that it is not possible to

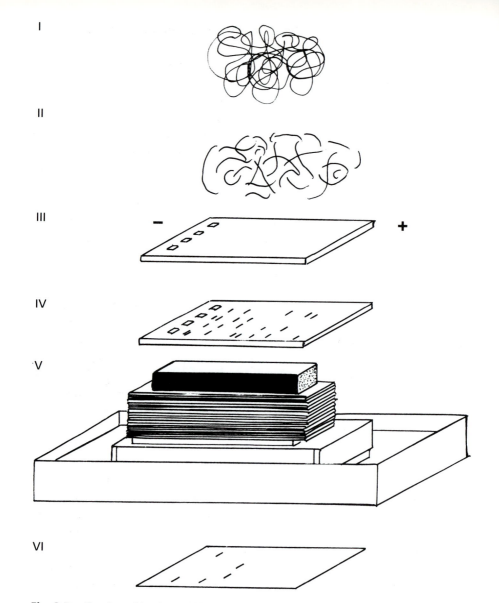

I

II

III

IV

V

VI

Fig. 3.5. Southern blotting and hybridisation. DNA to be tested is isolated (I), and digested by restriction endonucleases to produce numerous linear fragments of different sizes (II). These are loaded into the wells of an agarose gel (III), and separated electrophoretically (IV). The gel is then inverted onto a filter paper wick, a nitrocellulose or nylon filter is placed onto the gel, it is covered with filter paper and a pile of absorbent paper. The whole assembly is then weighted down with a heavy block (V). A buffer passes up the wick and through the gel, transferring DNA fragments onto the filter. The transferred DNA is then baked onto the filter in a single-stranded form, and reacted with a labelled probe. DNA hybrids are then detected (VI).

determine whether heteroduplex formation results from the presence of short sequences of absolute homology interspersed with longer sequences with little or no similarity, or from relatively poorly matched sequences that have sufficient similarity to stabilise the heteroduplex structure.

The ultimate test of the evolutionary relationship between organisms is a direct comparison of the nucleotide sequence of bacterial genes. Of particular interest in taxonomic studies are the genes encoding 16 S rRNA. Within a bacterial species this sequence is well conserved, but the sequence of the 16 S rRNA genes shows considerable divergence between species. The more divergent the nucleotide sequence of the genome is, the more distantly related are the organisms. The 16 S rRNA genes in bacteria are a good indicator of this divergence.

It is only with the development of nucleotide sequence studies that a proper phyletic classification of bacteria has become possible. Previously, bacterial taxonomy relied on a phenetic view of organisms. Until the 1920s bacterial taxonomy was in complete disarray, with many fragmentary classification systems, each devised by bacteriologists with interests only in specific areas of the subject. In 1923 in order to rectify this situation a committee, chaired by David Bergey, published a comprehensive taxonomic scheme including all of the then-known bacteria. This scheme has been periodically updated, but until the 1970s it represented an exclusively American view of bacterial taxonomy. In 1974 the venture became an international co-operative effort, with authors from across the globe making contributions to *Bergey's Manual of Determinative Bacteriology*. This has now been replaced by a four volume study: *Bergey's Manual of Systematic Bacteriology*. These volumes contain the internationally recognised names and descriptions of bacterial species. However, it should be borne in mind that bacterial taxonomy is steadily being refined, and that the interpretation of taxonomic data is constantly being updated. These changes are regularly communicated in the *International Journal of Systematic Bacteriology*.

Volume I of *Bergey's Manual* was published in 1984 and includes the Gram-negative bacteria. Volume II appeared in 1986 and describes the Gram-positive bacteria. Phototrophic and other specialised bacteria including gliding bacteria and archaebacteria are found in volume III, published in 1989. Volume IV was also published in 1989, and is devoted to actinomycetes and other filamentous bacteria.

3.5.3   The diversity of bacteria as represented by *Bergey's Manual of Systematic Bacteriology*

Some of the most devastating bacterial human pathogens are included in the first volume of *Bergey's Manual of Systematic Bacteriology*. Although devoted to Gram-negative bacteria this volume describes a vast array of microbes with a diverse biology. The majority of the bacteria described are harmless, and many are of benefit to humans. Those bacteria that are of benefit may be useful because of their natural biology, or may be harnessed for technological purposes. Many of the bacteria that comprise the internal commensal human flora are Gram-negative. We undoubtedly benefit from our commensal flora. The metabolic activity of microbes in the gut is equivalent to that of the liver. A few bacteria described in this section produce clinically useful antibiotics, and *Escherichia coli* is probably the most intensively studied of bacteria, largely because of its role in molecular genetics.

The diversity of bacteria described in this volume is a reflection of the comprehensive nature of Bergey's classification system: a strength that makes this system fundamental to bacteriology. Among the human pathogens described are the causative agents of diseases that have been recognised for centuries including plague, typhus, typhoid and cholera, and of diseases that have been newly recognised in human medicine such as legionnaire's disease, Lyme disease and diarrhoea caused by campylobacters. The first volume also includes important animal and plant pathogens. The sulphate- or sulphur-reducing bacteria cause economic damage because of their role in metal corrosion. They are particularly important in the corrosion of underground metal pipes. The genus *Bdellovibrio* comprises unusual Gram-negative bacteria that are parasites of other Gram-negative bacteria, living in the periplasmic space below the bacterial outer membrane.

Photobacterium phosphoreum represents a spectacular group of Gram-negative bacteria described in volume I of *Bergey's Manual*. Bacteria like this have developed a symbiotic relationship with a diversity of bony fish and also with members of the squid family (Cephalopoda). These bacteria live in specialised organs within the animals, where they produce light. This is a highly energy – demanding reaction that is catalysed by the enzyme **luciferase**.

The Gram-negative bacteria described in volume I of *Bergey's Manual* include bacteria such as mycoplasmas and the rickettsiae and chlamydiae that do not elaborate peptidoglycan as part of their cell walls. The rickettsiae and chlamydiae are obligate intracellular parasites. Volume I also includes a variety of obligately endosymbiotic bacteria that live within the cells of protozoa, fungi, insects and other invertebrates.

An anomaly of the Bergey classification system is that whereas most archaebacteria are described in volume III, a few are to be found in volume I together with the Gram-negative eubacteria. These include some halophiles and certain extreme thermophiles. In an entirely rational classification system, these bacteria would be included with the other archaebacteria.

The Gram-positive bacteria described in volume II are, in general, more nutritionally complex than the Gram-negative bacteria described in the first volume. There are two families of Gram-positive cocci: the Micrococcaceae and the Deinococcaceae. The latter family includes bacteria that are highly resistant to the damaging effects of radiation. These probably include the most radiation-resistant organisms on the planet. The endospore-forming Gram-positive rods are a diverse and important group of bacteria. Members of the genus *Bacillus* are aerobic bacteria, whereas the clostridia are obligate anaerobes. *Bacillus anthracis* causes anthrax, and members of the genus *Clostridium* are also pathogenic, causing diseases such as tetanus and gas gangrene. In each case, the ability of the causative agent to produce resistant spores adds to the pathogenicity of the bacterium. Members of the genus *Bacillus* are also of economic importance. Some produce antibiotics, others are used in molecular biology. As part of the sporulation process *Bacillus thuringiensis* produces an insecticide that is now used as a natural pest control. The regular, non-sporing, Gram-positive rods include lactobacilli, of importance in the food industry, and also incorporate the genus *Listeria*. This genus contains important pathogens of animals and humans. There is a large group of irregular Gram-positive rods, exemplified by the corynebacteria. These include important pathogens of humans, animals and plants, an example being *Corynebacterium diphtheriae*. This bacterium, if it carries an appropriate prophage, can cause diphtheria.

The nocardia are acid-fast, branching, non-sporing bacteria that are described in volume II along with the acid-alcohol-fast mycobacteria. The latter take up the Gram stain very poorly, and are classified here largely because of the architecture of their cell walls, rather than on their ability to stain using Gram's method. The nocardiform bacteria tend to grow as branching cells. The mycobacteria have plagued humans from time immemorial. *Mycobacterium tuberculosis*, as its name suggests, causes tuberculosis, and *Mycobacterium leprae* is the causative agent of leprosy, a disease described in the Old Testament. However, mycobacteria also play a crucial role in the biodegradation of industrial effluents, and are thus also of direct benefit to humans.

Volume III of *Bergey's Manual* includes the cyanobacteria (formerly known as blue-green algae) other photosynthetic bacteria and Gram-negative bacteria of a more complex nature than those described in volume I. The volume also contains descriptions of the major groups of archaebacteria, including some of the most ancient extant life forms. Cyanobacteria include highly differentiated photosynthetic prokaryotes that are important for their ability to fix atmospheric nitrogen. The enzyme nitrogenase responsible for this reaction is sensitive to oxygen, and so **nitrogen fixation** within cyanobacterial filaments takes place in specialised non-photosynthetic cells known as **heterocysts**. Budding, spinate, appendaged and sheathed bacteria are described in this volume. These bacteria tend to be found in aquatic habitats. The gliding bacteria are a diverse group of bacteria, the only common feature of which is their unusual gliding motility. The Myxobacteriales is an order of gliding bacteria that have the ability to form elaborate fruiting structures. They play an important role in the decomposition of plant material and animal dung.

The archaebacteria, described in volume III, are a very diverse group of bacteria, which differ in fundamental respects from other organisms. They are physiologically and genetically distinct from the eubacteria. Their cell walls are made of a variety of polymers, but do not contain peptidoglycan. *Bergey's Manual of Systematic Bacteriology* divides the archaebacteria into five groups. These are the methanogenic archaebacteria, the archaebacterial sulphate reducers, the extreme halophilic archaebacteria, the cell wall-less archaebacteria and the extremely thermophilic sulphur metabolisers. The archaebacteria can be found in some of the most extreme habitats on the planet.

The final volume of *Bergey's Manual* is devoted to the actinomycetes. All of the bacteria described in this volume were previously classified with the Gram-positive bacteria. They include bacteria with complex colonial morphologies having branching, aerial hyphae and an elaborate life-style. Some of the bacteria described in this volume produce spores. Many of the clinically useful antibiotics used today are produced by bacteria described in this volume.

3.5.4 Identification of bacteria

Once the characteristics of a group of bacteria have been determined, tests may be devised to distinguish individual species from one another. Very often the identification of bacteria relies upon the results of a very few key tests. In comparison with most higher organisms, bacteria are relatively simple structures. Paradoxically, this makes bacterial identification a relatively more difficult process than is the identification of higher organisms. This is a natural consequence of the observation that bacteria possess relatively few key features by which they may be identified. These include the following:

- *Morphology*: including cell size, shape and arrangement; the number and arrangement of flagella; possession of spores; their size, shape and location.
- *Chemical composition*: staining properties; possession of particular cytoplasmic inclusion bodies.
- *Metabolic characteristics*: possession of characteristic enzymes; the ability to attack specific substrates; production of unusual metabolic products.
- *Cultural characteristics*: nutritional requirements; the physical conditions required for growth, particularly temperature and atmosphere; colonial morphology; pigment production; characteristic smell; antibiotic resistance.
- *Pathogenicity*: *in vitro* tests for pathogenic features. (Historically these included attempts to reproduce disease processes in experimental animals. These tests are no longer practised.)
- *Antigenic structure*: particularly useful for differentiating strains at a subspecies level.
- *Genetic constitution*: the possession of plasmids and plasmid profiles; polymerase chain reaction-based typing methods.
- *Ecological characteristics*: the habitat of an organism may often provide useful clues to its identity.

In identifying bacteria, some or all of these features may be considered.

The microscopic examination of bacteria

The microscopic examination of bacteria, particularly when used in conjunction with a differential stain, permits much useful information to be obtained concerning the bacterium that is being observed. The Gram stain is the commonest of the differential strains. It yields information not only about the cell size and shape, it also permits inferences to be made about the bacterial chemical composition and structure. The Gram stain is the foundation stone of bacterial identification. In some circumstances examination of a Gram-stained smear from a clinical specimen, together with information on the source of the specimen, is all that is required for a provisional identification of the pathogen to be made. For example, Gram-positive diplococci found in the blood of a patient suffering from lobar pneumonia are almost certainly *Streptococcus pneumoniae*, and Gram-negative diplococci inside pus cells from a urethral discharge may be provisionally identified as *Neisseria gonorrhoeae*, the cause of gonorrhoea. However, bacteria with very different cultural characteristics and from different genera may be virtually indistinguishable when visualised microscopically.

The Ziehl Neelsen stain is another differential stain used in bacterial identification. It is used to identify acid-alcohol-fast bacteria, and is widely employed in the diagnosis of mycobacterial diseases. Since *Mycobacterium leprae* cannot be grown in artificial culture, the appearance of numerous acid-alcohol-fast bacilli in a specimen taken from a suspected leprous lesion is sufficient to confirm a diagnosis of leprosy. The presence of acid-alcohol-fast bacilli in sputum or in cerebrospinal fluid is usually sufficient to permit a provisional diagnosis of pulmonary tuberculous or tuberculosis meningitis. This is particularly important since it takes several weeks to grow and identify *Mycobacterium tuberculosis*, and a provisional diagnosis permits the prompt initiation of antituberculous therapy. The diagnosis of renal tuberculosis using a Ziehl Neelsen-stained specimen of urine is more difficult. The commensal bacterium *Mycobacterium smegmatis* grows on the external genitalia of many men and women. This bacterium is a relatively common contaminant of urine samples, and its presence may confuse the microscopic diagnosis of renal tuberculosis.

Flagellar stains were once important in taxonomic studies and in the identification of bacteria, but their use is now falling out of fashion. Volutin granules stain a reddish colour with methylene blue, and their presence may be used to identify corynebacteria. Spore stains are used to determine the size, shape and position of these structures within the sporangium. These features are key characteristics in the identification of members of the genera *Bacillus* and *Clostridium*.

Colony appearance and morphology

The appearance of bacterial colonies growing on the surface of an agar plate is of prime importance in the identification of bacterial cultures. Bacterial colonies display a diversity of size and colour that is characteristic of a particular strain. The size of colonies is independent of the magnitude of the cells that they contain, and colonies range from pinpoint size to dimensions sufficient to cover an entire plate. The colour, size and shape of colonies may show considerable variation when bacteria are grown on different solid media, even when the same strain is being cultured. Species of the genus *Proteus* grow as discrete buff-coloured colonies on MacConkey's agar, giving the typical appearance of non-lactose-fermenting members of the family Enterobacteriaceae. When *Proteus* sp. are grown on fresh blood agar, however, a single colony forming unit grows as a greyish mass and it swarms over the entire area of a Petri dish after overnight incubation. It is this characteristic that lends the genus its name. Proteus, the old man of the sea, had the ability to adopt different shapes according to Greek mythology. There is a variety of *Bacillus cereus*, var. *mycoides*, and its colonies show a characteristic swirling pattern with filaments all curling in one direction.

The surface of a colony may appear dry or glistening, rough or smooth. Some colonies appear to be mucoid. This characteristic is a feature of encapsulated bacteria. Repeated subculture on artificial medium can cause normally encapsulated bacteria to lose their capsules. In consequence of this loss, their colony morphology changes. Colonies may appear flat or convex, and may show irregularities of their surfaces. Colonies of *Streptococcus pneumoniae* often grow to resemble draughts pieces and sometimes are even called 'draughtsmen colonies'. A few bacteria have the ability to burrow into the surface of agar plates. An example is *Eikenella corrodens*, whose specific name reflects this habit. The outline form of a colony varies from circular, through an irregular shape, to filamentous colonies. Some bacterial colonies take on the appearance of tree roots, and these are described as rhizoid.

Bacterial colonies may also be pigmented. The names of some bacteria are derived from this feature. The genera *Flavobacterium* and *Xanthomonas* produce yellow colonies; *flavus* is the Latin and *xanthos* the Greek word for yellow. *Staphylococcus aureus* strains frequently produce golden colonies; *aurum* is the Latin for gold. *Melanos* is the Greek word meaning black and *Bacteroides melaninogenicus* grows as black colonies. *Chromobacterium violaceum* grows to yield violet or purple colonies. Environmental isolates of *Serratia marcescens* frequently produce **prodigiosin**, a deep red pigment. It has been proposed that the growth of pigmented strains of *Serratia marcescens* occurred on communion bread held in monastic safes. This may account for some of the mediaeval claims of miraculous transubstantiation: conversion of Eucharistic bread into flesh. The variety of pigments produced by bacteria is sufficient for microbiologists with artistic talents to paint coloured culture pictures on solid growth media.

Biochemical tests used in bacterial identification   The colony morphology and microscopic appearance of a bacterium and its Gram reaction having been determined, relatively few tests are required to identify bacteria to genus level. These include the relationship with oxygen, fermentation reactions and nitrogen metabolism. Other tests may be performed as appropriate. These include the production of hydrolytic enzymes, the nutritional requirements of bacteria, their antimicrobial sensitivity, expression of pathogenic factors, antigenic determinants and the genetic characteristics of the isolate.

Bacteria may have an absolute requirement for oxygen, or they may grow better in its presence than in its absence. Some bacteria are indifferent to the presence of oxygen and yet others are killed in its presence. Bacteria also have different metabolic responses to oxygen. These relationships can be usefully exploited in the identification of bacteria.

Glucose may be oxidised in the process of aerobic respiration. It may also be used fermentatively. The **Hugh and Liefson test** is used to determine whether glucose is oxidised or fermented. Two tubes containing a glucose medium together with a pH indicator are inoculated with a culture of the bacterium to be tested. One tube then has a layer of paraffin wax poured over the medium to exclude air from the culture. If glucose undergoes only oxidative metabolism, acid is produced only in the tube that is exposed to the air, and the pH indicator will change colour only in this tube. Alternatively, if the bacterium is capable of glucose fermentation, acid will be produced in the absence as well as the presence of air, and the pH indictor in both tubes will change colour. If the indicator in neither tube changes colour, this indicates that glucose is not metabolised.

For the oxidation of glucose many bacteria utilise a **respiratory transport chain,** a collection of **cytochromes** and other enzymes, terminating in cytochrome oxidase. Bacteria that produce cytochrome oxidase can oxidise tetramethyl-*para*-phenylene diamine hydrochloride, often referred to simply as TMPD or oxidase reagent, to produce an intensely coloured, purple product in less than 10 seconds.

Many bacteria that grow in air, with the notable exception of the streptococci, produce an enzyme called catalase that helps to protect them from toxic oxygen compounds. Catalase regulates the release of molecular oxygen from peroxides such as hydrogen peroxide. To test for catalase, a wide-bore capillary tube is half filled with hydrogen peroxide solution. The other end of the tube is used to collect a small portion of the bacterial colony to be tested. The hydrogen peroxide solution is then allowed to run down the tube and onto the bacterial cells. A positive result is indicated by the generation of oxygen bubbles. Caution must be observed when performing this test on colonies that have grown on medium containing fresh blood. If any medium is picked up with the bacterial cells this may give rise to a false positive result because blood contains catalase, and this will react in the same way as the bacterial enzyme.

Bacteria have the ability to attack a variety of carbohydrates. From these substrates they may produce acid alone, or acid together with gas. *Salmonella typhi* produces just acid from the fermentation of glucose, and can thus be easily distinguished from the other salmonellae that produce both acid and gas. Acid production is detected by growing bacteria in peptone water to which has been added the carbohydrate to be tested, together with a pH indi-

Fig. 3.6. Durham tube in a broth culture. When the broth culture is inoculated, the inverted Durham tube is filled with broth. Gas that is generated during incubation becomes trapped, forming a bubble, thus indicating gas production by microbes in the broth.

cator. Gas production is visualised with the aid of a **Durham tube**. This is a small inverted tube included within the liquid growth medium. At the point of inoculation of the sugar peptone water, the Durham tube is filled with the medium. If the test bacterium produces gas as a result of any metabolic process, the gas will collect in the inverted tube, and will appear as a bubble after appropriate incubation (Fig. 3.6).

Many bacteria produce organic acids from the metabolism of glucose. These may lower the pH of the growth medium to less than pH 4.4. This is sufficient to change the colour of the methyl red pH indicator to its red state. This forms the basis of the **methyl red (MR)** test. Other bacteria produce acetyl methyl carbinol, also known as **acetoin**, from glucose. This compound may be detected by adding potassium hydroxide to a glucose peptone broth culture of the bacterium under investigation. This forms a diacetyl product that can then react with the guanidine residues found in peptone, causing the development of a pink colour. This colour may be enhanced by the addition of 1-naphthol to the test reaction. This is the **Voges–Proskauer** or VP test. Bacteria that are positive for the VP test are usually negative for the MR test and vice versa since, in general, bacteria possess only one fermentation pathway. Exceptions include *Proteus mirabilis* and *Hafnia alvei*, which are positive for both the MR and the VP tests.

The metabolic products of bacteria may confer characteristic smells upon bacterial cultures. Some smells are pleasant; others are offensive. Anaerobic bacteria frequently generate putrid smells, but cultures of *Clostridium difficile* produce *para*-cresol. This confers an odour reminiscent of horse stables upon cultures of the bacterium. The aromatic products of anaerobic bacteria may be separated and identified by gas–liquid–chromatography. Individual species produce characteristic patterns of products, as seen by gas-liquid–chromatography, and these patterns may be used to identify anaerobic bacteria. Aerobic and facultative bacteria may also be associated with a characteristic odour, but these smells are not used in the definitive identification of these bacteria. Rather, they are used as a guide to the probable identity of an isolate.

The ways in which bacteria metabolise nitrogen can also be exploited for identification tests. The production of **indole** from tryptophan indicates that bacteria possess an enzyme complex known as **tryptophanase**. Kovac's reagent, *para*-dimethyl-aminobenzaldehyde dissolved in hydrochloric acid and isoamyl alcohol, condenses with the **pyrrole ring** structure in indole to produce a quinoidal product that has an intense red colour.

The ability to split urea to liberate water, carbon dioxide and ammonia is found in many bacteria. The action of urease causes urea-based medium to become alkaline. The production of urease is demonstrated by growing the bacterium to be tested in urea broth containing a pH indicator.

Bacteria may reduce nitrates to nitrites and some may further reduce nitrites to ammonia. Addition of α-naphthylamine to a nitrate broth culture will cause the production of a red diazo-complex if the bacterium being tested has reduced the nitrate in the medium to nitrite. If a red diazo-complex does not form, zinc dust is added to the culture. The zinc dust will reduce any nitrate remaining in the medium to nitrite, and this may then form a diazo compound. If a red colour forms following addition of the zinc dust, this shows that the medium still contained nitrate, indicating that the bacterium being tested could not reduce nitrates. If, however, the bacterium can reduce nitrites as well as nitrates, a red colour will not form upon the addition of zinc dust to the medium, and the test may be regarded as positive.

Bacteria may produce hydrolytic enzymes that alter the medium upon which they grow. These include **proteases, haemolysins, DNase, amylase, and lipases**. Proteases may attack different proteins and produce a variety of effects in various media. On heated blood agar they may cause an alteration in the chocolate colour, whereas on milk agar, casein agar or egg yolk agar plates they may cause the appearance of a clear halo around protease-producing colonies. Gelatin hydrolysis may be detected by a clear halo around colonies after precipitation of any residual gelatin in the medium upon the addition of 10% potassium dichromate in hydrochloric acid.

Haemolysis is detected by culturing bacteria on fresh blood agar plates. Haemolysins cause the breakdown of red blood cell membranes and cellular contents. Streptococci may produce a partial haemolysis, causing a greenish colour to appear within the fresh blood agar plate. This is called α-haemolysis. Complete haemolysis may also be accomplished by other streptococci and this is known as β-haemolysis. It is desirable to reserve the terms α- and β-haemolysis to describe the streptococci. Some haemolysins are oxygen-sensitive, and haemolysis may be enhanced by anaerobic incubation of the test culture.

The production of DNase may be demonstrated by growing bacteria on an agar plate that contains DNA. Following incubation, the remaining DNA is precipitated by flooding the plate with hydrochloric acid. If a clear halo appears around colonies, this indicates that the bacterium produces DNase. Similarly amylase production may be detected by growth of bacteria on a starch-based medium. After the culture has grown the plate is flooded with a dilute iodine solution. Absence of a dark blue starch-iodine complex around the colonies indicates amylase production.

Lipases may be detected using egg yolks. Lecithinase (phospholipase C) causes a dense precipitate to form in the medium as a result of the hydrolysis of lecithin. Other lipases cause the formation of an iridescent scum on the surface of egg yolk plates.

Nutritional requirements of bacteria   The nutritional demands of fastidious bacteria may be exploited in their identification. The human pathogen *Haemophilus influenzae* requires two growth factors known as the **X** and **V** **factors**. The X factor is identified as haem, and the V factor is **nicotinamide adenine dinucleotide**, often referred to as **NAD**. By seeding a nutrient agar plate with *Haemophilus influenzae* and placing filter paper discs on the lawn, growth will occur only around the disc containing both the X and V factors, and not around discs containing the X factor or the V factor alone. In contrast, the human commensal bacterium *Haemophilus parainfluenzae* requires only the V factor since it can elaborate its own haem compounds. This bacterium will grow around both the V disc and the X + V disc. These bacteria also show **satellitism**. This is seen as enhanced growth around colonies of *Staphylococcus aureus* growing on fresh blood agar. The fresh blood supplies the haem, and the *Staphylococcus aureus* provides an excellent source of NAD.

Antimicrobial sensitivities   Some bacteria display specific sensitivity or resistance to antibiotics. *Staphylococcus saprophyticus* is differentiated from other staphylococci because it is resistant to novobiocin. *Streptococcus pyogenes* is sensitive to bacitracin, whereas other β-haemolytic streptococci are resistant to this antibiotic. Similarly *Streptococcus pneumoniae* can be differentiated from other α-haemolytic streptococci because it is sensitive to optochin, whereas the others are resistant to this agent. *Streptococcus pneumoniae* is also differentiated from other α-haemolytic streptococci because it lyses in the presence of a weak solution of bile salts. Bacteriocin sensitivity can be used to differentiate bacterial strains at a subspecies level.

Pathogenicity factors   Pathogenic bacteria elaborate products that assist in their ability to cause disease. These are their pathogenicity factors. Tests have been developed to detect pathogenicity factors. *Staphylococcus aureus* produces **coagulase**, an enzyme that causes the clotting of blood plasma. If a culture of *Staphylococcus aureus* is mixed with citrated plasma in a tube and incubated at 37 °C, a clot will form. Coagulase may also be detected as a clumping factor. A suspension of *Staphylococcus aureus* in saline is mixed with a loopful of plasma on a glass microscope slide. After mixing, coagulase causes the formation of **fibrin** strands. In turn, these strands of fibrin cause the bacteria to clump together.

One of the principal toxins of *Clostridium perfringens*, the causative agent of gas gangrene, is a lecithinase, also called α-toxin. This is detected using the **Nagler test**. Half an egg yolk plate is covered with **antitoxin** and an inoculum of the test organism, together with positive and negative controls, is inoculated from the side of the plate that is not covered in antitoxin across to the side that did receive the antitoxin. After incubation α-toxin will cause precipitation in the medium around the **toxigenic** bacteria on the side of the plate that was not covered in antitoxin, but the specific antitoxin prevents the formation of the precipitate on the other side of the plate.

The ability of *Corynebacterium diphtheriae* to produce disease depends upon the possession of a prophage that encodes toxin production. Only strains that carry the prophage can elaborate the diphtheria toxin, and thus only these strains cause disease. Toxin production may be demonstrated using the **Elek test**. Control and test strains are streaked across an agar plate, and a filter paper strip soaked in antitoxin is placed across the inocula. If the test strain produces toxin, the toxin will diffuse out from the streak. Likewise, the antitoxin diffuses out from the filter paper strip. Where the toxin and the antitoxin meet, a precipitate is formed within the agar. This demonstrates the ability of the strain to produce toxin, and hence cause disease.

The antigenic structure of bacteria   The antigenic structure of bacteria can be usefully exploited in their identification and in discriminating between strains at a subspecies level. Salmonellae can be subdivided into about 2000 **serovars**, depending upon the pattern of expression of their somatic 'O' antigens and the flagellar 'H' antigens. **Serotyping** may be applied to many bacteria. Antisera may be **polyvalent**, reacting with several antigens or **monovalent** reacting with only one antigen.

Immunological typing of bacteria has been considerably enhanced by the development of **monoclonal antibodies**. However, bacteria of different species may express antigens that cross-react with the same antibody. Bacterial antigens may even cross-react with antigens on the surface of animal or human cells.

Genetic constitution of bacteria   Strains of many bacteria may be identified to subspecies level by examination of their plasmid content. Plasmids may be extracted by alkaline lysis of cells, and then separated according to their size on agarose gels. Different strains of bacteria display different plasmid profiles.

The use of gene probes has also been exploited in the identification of bacteria. Bacteria possess species- or strain-specific genes. The DNA encoding such genes may be labelled and used in hybridisation experiments to identify isolates carrying the sequence complementary to that of the probe DNA. The earliest gene probes comprised relatively long fragments of DNA, and these sometimes suffered from the problem of non-specific hybridisation. This results from binding of the probe to unrelated DNA sequences. Such reactions lead to false-positive identifications. Probes have been refined and sequences as short as 10–25 nucleotides are now employed in hybridisation studies. These short sequences are known as **oligonucleotide probes**, and have proved to be much more specific than longer DNA fragments.

A further development in bacterial identification by exploitation of their genetic constitution has been the application of the **polymerase chain reaction** (PCR) to the identification of bacteria and other microbes, especially non-culturable viruses. PCR involves the use of a pair of oligonucleotide sequences that act as primers for DNA replication. The first primer is designed to **anneal** to one DNA strand at one end of the target sequence. The second primer is designed to anneal to the complementary DNA strand at the other extreme of the DNA fragment to be amplified. These primer pairs are used to replicate the target DNA in conjunction with a DNA polymerase isolated from a thermophilic bacterium, *Thermus aquaticus*. This is achieved by melting the target DNA, typically at temperatures of about 95 °C, in the presence of both primers. The mixture is then cooled to allow annealing of the primers to the target DNA. The annealing temperature can vary considerably because of the nucleotide sequence of the oligonucleotide primers, but is frequently below 65 °C. The temperature of the reaction mixture is then raised to 72 °C to allow replication of the target DNA, mediated by the DNA polymerase. After a period of replication, the reaction temperature is raised once more to melt the products of replication. Re-cooling permits the annealing of primers to the targets again, and this is followed by another round of DNA replication. After 20 or so cycles, the original target has been replicated many million-fold. This generates a DNA fragment of a predictable size. Because of the specificity of binding of oligonucleotide primers, PCR technology permits highly specific identification of DNA sequences. PCR technology is particularly useful in the identification of bacteria that are difficult to cultivate, e.g. *Borrelia burgdorferi*, the cause of Lyme disease, or of slow-growing bacteria such as *Mycobacterium tuberculosis*.

PCR technology can also be used for the subtyping of bacteria. The genes encoding flagellin proteins found in bacterial flagella in *Campylobacter jejuni* are highly variable. This is the basis of 'H' antigen variability. Fortunately, the variable region of the gene is flanked by two constant regions. PCR products generated from flagellin genes, amplified using primers that bind to each of the constant regions, may be cut with **restriction endonucleases**. These are enzymes that recognise specific DNA sequences and cut the DNA at these locations. DNA fragments generated by restriction endonuclease digestion may be separated by size using gel electrophoresis to give a DNA fingerprint. The flagellin genes in bacteria of different 'H'-types give PCR products that yield different restriction endonuclease digest fingerprints.

PCR technology is a new tool for bacteriologists, as is the application of monoclonal antibodies. These two methods are now being combined to assist in the identification of bacteria. Monoclonal antibodies are attached to magnetic beads, and the resultant immunomagnetic beads are mixed with samples containing different types of bacterium. The monoclonal antibodies will bind only to the target bacterium, which may then be isolated from the mixture magnetically. The purified organisms may be present in very small numbers. Even so, there will be sufficient to permit detection using PCR technology with primers that are specific for the target bacterium. Such techniques may become important in the diagnosis of infectious diseases caused by organisms that are difficult to culture. PCR technology is also being used to classify bacteria that cannot be grown in artificial medium. This is accomplished by amplifying part of the 16 S rRNA gene that is flanked by conserved sequences, and determining the nucleotide sequence of the amplified products. In this way the degree of relatedness of unculturable organisms is being assessed.

3.6   Virus classification

Early attempts to classify animal viruses were based upon their common pathogenic properties. Hepatitis viruses, for example, were classified under such schemes and included a small RNA virus, known as hepatitis A virus, and a DNA virus, known as hepatitis B virus. These viruses were classified along with the yellow fever viruses. The only common feature that these viruses share is a common **organ tropism**. Similarly, respiratory viruses were classed together and these include influenza virus, rhinoviruses and adenoviruses. Common ecological characteristics were also used to group viruses. The arthropod-borne viruses, also called arboviruses, include members that are now considered to belong to several distinct groupings. Such classification schemes, like similar schemes used for bacterial classification, were ultimately found to be inadequate.

To address these difficulties the International Committee on Nomenclature of Viruses, or ICNV, was established in 1966 at the International Congress of Microbiology held in Moscow. At that time several hundreds of new viruses were being isolated and characterised, and there was considerable debate concerning how they should be classified and assigned to taxonomic groups. As many virus characteristics as were then conceivable were considered, and weighted with respect to their use as criteria for dividing groups of viruses. A system of classification was devised that did not involve any hierarchical levels higher than the family. The scheme also made no attempt to include phylogenetic relationships. The present system for virus identification thus includes families, genera and species. Lower levels of classification, including subspecies, strains and variants are used by specialist laboratories and virus culture collections. Occasionally subfamilies may be included as part of the division.

The criteria used for classification of viruses into families include nucleic acid type and structure and the morphology of the virion as visualised by electron microscopy. This may include such features as the capsid symmetry, number of capsomeres present and the absence or presence of an envelope, together with the nature of this structure. Classification of a virus at the genus level may include serological considerations, the strategy that the virus uses for replication and the method of transmission of the virus. Species divisions rely on such characteristics as **restriction endonuclease polymorphism**, and nucleotide sequence analysis. Restriction endonucleases are enzymes that recognise and cut DNA at specific sequences. **Restriction endonuclease polymorphism** describes the different patterns that related DNA sequences produce after digestion with restriction endonucleases, and separation of the digestion products by agarose gel electrophoresis. A taxonomic chart of selected viruses is shown in Table 3.2.

Table 3.2 *Classification of viruses*

Family	Characteristics	Typical members	Diseases caused (or uses)
Poxviridae	Double-stranded DNA; 'brick-shaped' particle; largest viruses	Vaccinia	Smallpox vaccine vector for virus genes
Herpesviridae	Double-stranded DNA; icosahedral capsid; enveloped;latency in the host is common	Herpes simplex	Cold sores or genital infections
		Varicella-zoster	Chicken pox
		Cytomegalovirus	Foetal retardation, pneumonia
		Epstein–Barr virus	Infectious mononucleosis (glandular fever)
Adenoviridae	Double-stranded DNA; icosahedral capsid with extending fibres; non-enveloped	Adenovirus (many types)	Respiratory and eye infections; tumours in experimental animals
Papovaviridae	Double-stranded circular DNA; 72 capsomeres in capsid; non-enveloped	Human papillomavirus	Warts and verrucas
		JC virus	Progressive multifocal leukoencephalo-pathy
		BK virus	Kidney problems
Hepadnaviridae	One complete DNA minus strand with a 5′ terminal protein;DNA is circularised by an incomplete plus strand; 42 nm enveloped particle	Hepatitis B virus	Hepatitis. There is also an association with carcinoma of the liver
Paramyxoviridae	Single-stranded RNA; virus particles are enveloped, and carry spikes	Parainfluenza	Croup
		Measles	Measles
		Respiratory syncytial virus	Bronchiolitis

Table 3.2 (*cont.*)

Family	Characteristics	Typical members	Diseases caused (or uses)
Orthomyxoviridae	8 segments of single-stranded RNA; helical nucleocapsid; virus enveloped, with spikes	Influenza	Influenza
Reoviridae	10–12 segments double-stranded RNA; icosahedral; non-enveloped	Rotavirus	Infantile diarrhoea
Picornavirdae	Single-stranded RNA; 22 to 33nm particles have cubic symmetry; non-enveloped	Poliovirus Coxsackie virus Rhinovirus Hepatitis A	Poliomyelitis Bornholm disease and infections of the heart Common colds Infectious hepatitis
Togaviridae	Single-stranded RNA; icosahedral nucleocapsid within an envelope particle	Rubella virus	German measles
Rhabdoviridae	Single-stranded RNA; bullet-shaped; enveloped particle	Rabies virus	Rabies
Retroviridae	Single-stranded diploid RNA; enveloped particles with icosahedral nucleocapsids; reverse transcriptase activity	Feline leukaemia virus Human immuno-deficiency virus	Leukaemia in cats AIDS

A further two examples of virus classification are presented in Table 3.3.

Table 3.3 *Classification of virus families*

Family	**Herpesvirdae** Double-stranded DNA of M_r 80–150. The virus particle is a 130 nm icosahedron enclosed in a lipid envelope. The virus buds from the nuclear membrane of the host cell. Most herpesviruses can exist in their host in a latent state
Subfamily	**Alphaherpesvirinae** (the herpes simplex-like viruses)
Genus	**Simplexvirus** (herpes simplex-like viruses) **Varicellavirus** (varicella and pseudorabies-like virus)
Subfamily	**Betaherpesvirinae** (the cytomegaloviruses)
Genus	**Cytomegalovirus** (the human cytomegalovirus)
Genus	**Muromegalovirus** (the mouse cytomegalovirus)
Subfamily	**Gammaherpesvirinae** (the lymphocyte-associated viruses)
Genus	**Lymphocryptovirus** (Epstein–Barr-like viruses)
Genus	**Rhadinovirus** (saimiri–ateles-like viruses)
Family	**Paramyxoviridae** Single-stranded RNA of M_r 5–7. The viruses are enveloped 150 nm particles that carry spikes. The nucleocapsids are 12–17 nm in diameter and contain the enzyme RNA transcriptase. Filamentous forms are common. Cytoplasmic budding occurs from the host cell membrane
Genus	**Paramyxovirus** (parainfluenza)
Genus	**Morbillivirus** (measles-like viruses)
Genus	**Pneumovirus** (respiratory syncytial virus)

It is beyond the scope of a book of this size to show all virus families. The chosen examples of virus classification, given purely for reference purposes and chosen at random, demonstrate the division of viruses into genera.

Further reading

The interested reader may wish to consult the following texts in their future studies.

Barnett, J.A., Payne, R.W. & Yarrow, D. *The yeasts*, 2nd edition (1990) Cambridge University Press.

Bergey's manual of systematic bacteriology, volume I, Kreig, N.R., ed. (1984); volume II, Sneath, P.H.A., ed. (1986); volume III, Stanley, J.T., ed. (1989); volume IV, Williams, S.T., ed. (1989) Williams and Wilkins.

Brock, T.D., Smith, D.W. & Madigan, M.T. *The biology of micro-organisms*, 7th edition (1991) Prentice-Hall.

Carlile, M.J. & Watkinson, S.C. *The fungi* (1994) Academic Press.

Dale, J.W. *Molecular genetics of bacteria*, 2nd edition (1994) John Wiley & Sons.

de Duve, C. *A guided tour of the living cell* (1985) Freeman.

Deacon, J.W. *Introduction to modern mycology*, 2nd edition (1984) Blackwell Scientific Publications.

Department of health code of practice for the prevention of infection in clinical laboratories and post-mortem rooms (1978) Her Majesty's Stationary Office.

Dimmock, N.J. & Primrose, S.B. *Introduction to modern virology*, 4th edition (1994) Blackwell Scientific Publications.

Gerhardt, P., Murray, R.G.E., Costilow, R.N., Nester, E.W., Wood, W.A., Krieg, N.R. & Phillips, G.B. (eds.) *Manual methods for general and molecular bacteriology* (1994) American Society for General Microbiology.

Gow, N.A.R. & Gadd, G.M. (eds.) *The growing fungus* (1994) Chapman & Hall.

Margulis, L. & Schwartz, K.V. *Five kingdoms, an illustrated guide to the phyla of life on earth* (1982) Freeman.

Nester, E.W., Roberts, C.E. & Nester, M.T. *Microbiology: a human perspective* (1995) William C. Brown.

Postgate, J. *Microbes and man*, 3rd edition (1992) Cambridge University Press.

Prescott, L.M., Harley, J.P. & Klein, D.A. *Microbiology*, 2nd edition (1993) William C. Brown.

Rij, K.V. *The yeasts – a taxonomic study* (1984) Elsevier Science Publishers.

Rogers, H.J. *Bacterial cell structure* (1983) van Nostrand Reinhold.

Russell, A.D., Hugo, W.B. & Ayliffe, G.A.J. *Principles and practice of disinfection* (1992) Blackwell Scientific Publications.

Schlegel, H.G. *General microbiology*, 7th edition (1993) Cambridge University Press.

Tempest, D.W. Dynamics of microbial growth. In *Essays in microbiology* (Norris, J.R. & Richmond, M.H., eds.) (1978) John Wiley & Sons.

Tribe, M.A., Eraut, M.R. & Snook, R.K. *Light microscopy* (1975) Cambridge University Press.

Voyles, B.A. *The biology of viruses* (1993) Mosby.

Watson, J.D., Hopkins, N.H., Roberts, J.W., Steitz, J.A. & Weiner, A.M. *Molecular biology of the gene*, 4th edition (1987) Benjamin/Cummings.

Whittenbury, R. Bacterial nutrition. In *Essays in microbiology* (Norris, J.R. & Richmond, M.H., eds.) (1978) John Wiley & Sons.

Woese, C.R. Bacterial evolution. *Microbiological Reviews* (1987) **51**, 221–71.

Glossary

Abscess   Localised collection of pus in the body.

Acervulus   Flat or saucer-shaped bed on which conidiophores are produced.

Acetoin   Acetyl methyl carbinol: a fermentation product of glucose made by certain bacteria, and detected in the Voges–Proskauer test.

Acid-alcohol-fast   The description of structures that do not lose strong carbol fuchsin stain when treated with alcohol containing a mineral acid. This is a property of mycobacteria.

Acid-fast   The description of structures that do not lose strong carbol fuchsin stain when treated with mineral acids. This is a property of mycobacteria and nocardiform bacteria.

Acidophile   An organism that grows optimally at an acid pH.

Aerobe   An organism that requires oxygen for growth.

Aerosol   A dispersion of fine solid or liquid particles in air. Aerosols are easily created and dispersed when liquids are agitated, and may contain infectious particles if an inoculated culture is shaken.

Aerotolerant anaerobe   An organism that grows as well in the absence of oxygen as in its presence.

Aflatoxin   Carcinogenic toxin produced mainly by strains of the mould *Aspergillus flavus*.

Agar   Polysaccharide extract of seaweed used as a gelling agent in microbiological media. Agarose is a highly purified form of agar used for electrophoresis.

AIDS   Acquired immune deficiency syndrome, a condition resulting from infection with the human immunodeficiency virus, and characterised by the loss of a normal immune response. AIDS patients suffer from opportunistic infections, fever, weight loss and swollen lymph glands. They are also more prone to developing certain cancers than others.

Alkalinophiles   An organism that grows optimally at alkaline pH.

Allantoic cavity   A cavity in an embryonated egg that is filled with allantoic fluid. Allantoic fluid can be used for cultivating viruses such as the influenza virus. The allantoic cavity is enclosed within the allantoic membrane.

Allergen   A substance that elicits an allergic response in susceptible individuals.

Allergy   A hypersensitivity to the presence of certain antigens known as allergens, characterised by a pathological response in the presence of the allergen. Allergic responses are frequently manifest as skin rashes or asthma, but in severe cases may result in life-threatening anaphylactic shock.

Amnion   The innermost membrane surrounding the foetus.

Amphitrichous   Having a single flagellum at each end of the cell.

Amylase   An enzyme that breaks down starch.

Anaerobe   An organism that does not grow in the presence of air. See *obligate, facultative* and *aerotolerant anaerobe*.

Anamorph   Asexual phase of a fungus.

Anastomosis   (Plural: anastomoses.) Literally to furnish with a mouth. Used to denote cross-connecting hyphae, arteries, etc.

Animalcule   A microscopic animal.

Annealing (of DNA)   The formation of a double-stranded structure by the joining of complementary single-stranded molecules of DNA.

Antheridium   (Plural: antheridia.) A male gametangium.

Antibiotic   A substance that is produced by a microorganism that in very low concentration inhibits or kills the growth of another microorganism. Antimicrobial agents include synthetic compounds that have the same effect as antibiotics.

Antibody   A type of protein produced by vertebrate animals in response to the presence of a foreign antigen, and that binds in a highly specific manner to that antigen.

Antigen   Substance that elicits the production of antibodies. Antigens are usually carbohydrates or of a proteinaceous nature.

Anti-receptor   A protein or glycoprotein on the surface of a virus that attaches to a receptor site on a potential host cell during the initiation of virus infection of a cell.

Antiseptic   A chemical that is applied to the body surfaces to prevent infection. This is achieved by inhibiting or killing microorganisms.

Antiserum   (Plural: antisera.) Serum derived from blood that contains particular antibodies.

Antitoxin   An antibody that can bind to a toxin to prevent it from causing damage.

Aplanospore   A non-motile spore such as those produced by many primitive terrestrial fungi.

Apothecium   (Plural: apothecia.) An open-structured ascocarp produced by some ascomycete fungi.

Arabinogalactan   Polymer of arabinose and galactose that is covalently linked to the peptidoglycan of coryneform bacteria.

Archaebacteria   A major group of bacteria, some of which live in extreme environments, and that diverged from the eubacteria at a very early stage of evolution.

Arthric   See *holoarthric*.

Arthrospore   An asexual spore formed by the fragmentation of a fungal mycelium. These structures are also known as **arthroconidia**.

Ascocarp   A fruiting body of ascomycete fungi also called an **ascoma**. There are three main types of ascocarp, the cleistothecium, the perithecium and the apothecium.

Ascogenous hypha   (Plural: ascogenous hyphae.) Dikaryotic cell formed by the fusion of gametangia in the sexual reproduction of ascomycete fungi.

Ascogonium   (Plural: ascogonia.) Female gametangium formed by ascomycete fungi.

Ascomycetes   Fungi that have sac-like asci as their reproductive structures. Also known as **Ascomycotina**

Ascospore   Sexual spore formed within an ascus by ascomycete fungi.

Ascus   (Plural: asci.) A sac-like structure formed by ascomycete fungi in which ascospores develop.

Asexual reproduction   Reproduction that does not involve the fusion of gametes (sex cells).

Attenuated viruses   Viruses that have been treated so that they lose the ability to cause overt disease, but that retain the ability to infect cells.

Attenuation   Weakening of virulence.

Autoradiography   Technique used to localise a radioactive probe by its capacity to fog photographic film.

Autotroph   An organism that can assemble all its organic components from inorganic matter. Literally, a self-feeding organism.

Auxotrophic mutant   A mutant that requires growth supplements that are not needed for the growth of wild-type strains.

Bacillus   (Plural: bacilli.) A rod-shaped bacterium. The term bacillus, written with a lower case initial, should not be confused with the genus *Bacillus*, a group of aerobic, sporing Gram-positive rods.

Bactericidal agent   An agent that can kill bacteria.

Bacteriocin   A protein produced by strains of bacteria to kill other strains of the same, or closely related, species.

Bacteriophage   A virus that infects bacteria.

Bacteriostatic agent   An agent that prevents bacterial growth.

Bacterium   (Plural: bacteria.) A prokaryotic microorganism.

Bacteroid   Highly degenerate and irregular form adopted by bacteria of the genus *Rhizobium* when growing in root nodules of nitrogen-fixing plants.

Basal salts medium   A solution that provides an isotonic environment in which cells may be maintained for a short period. Basal salts media form the basis of tissue culture growth media.

Basidiocarp   Fruiting body formed by basidiomycete fungi.

Basdiomycetes   Fungi that can elaborate complex and conspicuous fruiting bodies. Also known as **Basidiomycotina.**

Basidiospore   A sexual spore formed by the basidium of basidiomycete fungi.

Basidium   (Plural: basidia.) A sexual structure that supports the basidiospores of basidiomycete fungi.

Binary fission   Splitting into two parts. Binary fission of bacterial cells is the means by which most bacterial populations grow.

Bioluminescence   The production of light by living organisms.

Biomass   The mass of living matter present in a specified location.

Biosynthesis   The formation of complex organic molecules from simple precursors, requiring energy input.

Blastic conidium   A conidium that forms by budding or swelling out of fertile hyphae and cut off by a septum.

Blastospore   A spore that is produced by budding or by swelling.

Boil   Inflamed swelling containing pus formed by the infection of a hair follicle.

Botulism   Clinical condition that is often fatal, characterised by flaccid paralysis and caused by the toxin of *Clostridium botulinum.*

Budding   A form of asexual reproduction, seen typically in yeasts, where the new cell grows by swelling from its parent.
Also: the passing of a virus capsid through the nuclear or plasma membrane during its acquisition of a virus envelope.

Burst size   The number of bacteriophage particles released when an infected bacterium lyses.

Canine distemper   A virus infection of dogs that causes a cough, weakness and catarrh.

Capsid   The protective protein coat that surrounds virus nucleic acid.

Capsid viruses   Viruses that lack envelopes.

Capsomere   A small structural unit within the virus capsid. Capsomeres may contain a single protein or several different proteins.

Capsule   An envelope of carbohydrate or protein surrounding the cell wall of certain microorganisms. The presence of a capsule may increase the virulence of a microorganism.

Carbuncle   A group of neighbouring hair follicles that become infected with *Staphylococcus aureus*. Carbuncles are inflamed and swollen, and have numerous heads that ooze pus.

Carcinogen   Compound that induces cancer.

Carcinoma   A malignant growth derived from epithelial cells.

Cardinal temperatures   The three temperatures that define the growth of a microorganism. They are the **minimum growth temperature, the optimum growth temperature** and the **maximum growth temperature**.

Casein   Protein derived from milk.

Catalase   Enzyme that breaks down hydrogen peroxide to release water and oxygen.

Catarrh   Inflammation of mucous membranes resulting in the overproduction of mucus.

Catheter   A tube that passes into the body to assist in the flow of fluids.

Cationic   Positively electrostatically charged.

CD4 marker   A receptor on the surface of lymphocytes that interacts with antigens. In this case 'CD' stands for cluster of differentiation, and over 70 such markers have now been described.

cDNA   DNA that is made from an RNA template using reverse transcriptase.

Cell monolayer   A layer of cells in a tissue culture that forms on the glass or plastic surface of the culture vessel, and that is only one cell thick.

Cell passage   The re-seeding of a sample of tissue culture cells in fresh tissue culture medium.

Cell transformation   Biochemical changes that occur within a eukaryotic cell that alter its morphology and behaviour. Many transformed cells become capable of tumour production. The changes within cells are at the level of the DNA and may be caused by viruses.

Cell tropism   The propensity for particular viruses to infect certain cell types. Cell tropism results from the presence of specific receptor sites on the surface of target cells that interact with anti-receptors on the surface of the virus.

Cell wall   A rigid structure, external to the cell membrane, that lends structure to plant, fungal and bacterial cells.

Cellulose   A polymer of glucose that forms the structural material of plant cell walls and some lower fungi.

Centriole   A subcellular organelle found in many eukaryotic cells, and involved in cell division.

Cerebrospinal fluid   A watery fluid that bathes the brain and spinal cord, and is also found within the ventricles of the brain. It is commonly referred to as CSF.

Chemoautotroph   An organism that obtains its energy from the metabolism of inorganic compounds, and that uses carbon dioxide as a sole carbon source. Chemoautotrophs do not need a supply of organic matter to grow.

Chemolithotroph   An organism that oxidises inorganic compounds to derive energy, and that uses carbon dioxide as its sole carbon source.

Chemoorganotroph   An organism that uses organic compounds as a source of energy, and that does not require light to grow.

Chemostat   A continuous flow culture apparatus in which fresh nutrients are added at the same rate as spent medium is removed from the vessel.

Chemotaxis   Movement in response to a chemical stimulus.

Chemotroph   An organism that uses chemical compounds for its energy supply.

Chitin   A polymer of N-acetylglucosamine that is present in the cell wall material of fungi, and in the exoskeletons of arthropods such as insects.

Chlamydospore   A thick-walled, resistant spore formed from a fungal mycelium. These structures are also known as **chlamydoconidia**.

Chloroplast   A photosynthetic organelle found in certain plant cells. Chloroplasts are a type of plastid that contain the light-harvesting pigment chlorophyll.

Cholera   Life-threatening, water-borne infectious diarrhoea caused by the bacterium *Vibrio cholerae*.

Chorioallantoic membrane   The membrane in which the allantoic membrane fuses with the chorion, an embryonic membrane that encloses all of the foetal tissues.

Chromatophore   A pigment-bearing body. This term is often applied to pigmented membranes found in bacteria.

Chromosomes   DNA bodies that contain all of the genetic information necessary for the continuation of the cell. The number of chromosomes per nucleus in a eukaryotic cell is constant for each species. Not all the genetic material of the cell is carried on chromosomes. Mitochondria and chloroplasts contain their own DNA, and many bacterial cells carry plasmids.

Cidal   Capable of killing.

Cilium   (Plural: cilia.) A hair-like appendage to a cell that is capable of movement. With ciliate microorganisms, cilia may be used as a means of propulsion.

Clamp formation   The process whereby when the binucleate dikaryotic cells of basidiomycete fungi divide, one pair each of nuclei goes into each daughter cell.

Cleistothecium   (Plural: cleistothecia.) An ascocarp, or fruiting body formed by ascomycete fungi that completely encloses the asci that it contains.

Clone   A collection of organisms derived from a single individual, and with a common genotype. It should be noted that because of spontaneous mutations, all the members of a clone do not carry identical copies of all of their genes. To clone is artificially to generate multiple copies of a gene or genes of an organism.

Cloning vector   A specialised small plasmid, incapable of self-mobilisation, that is used to introduce foreign DNA into a bacterial cell.

Coagulase   An enzyme that clots plasma by causing the formation of fibrin from fibrinogen. This enzyme is characteristically produced by *Staphylococcus aureus*.

Coccobacillus   (Plural: coccobacilli.) Bacteria that exist as such short rods (bacilli) that they appear as almost round (coccal).

Coccus   (Plural: cocci.) A round-shaped bacterium. Some cocci are markedly deformed.

Coenocytic hypha   A hypha that contains many nuclei, but no regular septa. Such hyphae are characteristic of lower fungi.

Cold sores   Vesicular eruptions around the mouth, resulting from re-activation of a latent herpes simplex virus infection. Cold sores may be triggered by many stimuli including stress, exposure to sunlight and menstruation.

Colony forming unit   The cell or cluster of cells that gives rise to a single colony when plated onto a solid medium.

Columella   The dome-shaped apex of the sporangiophore in certain phycomycete fungi.

Commensal flora   The normal microbial flora found in association with higher organisms.

Conidium   (Plural: conidia.) Also known as a **conidiospore**, this is an asexual spore that is borne externally by fungi.

Conidiophore   A specialised hyphal structure that supports conidia.

Conjugation   A mating process in which donor and recipient cells are temporarily joined to facilitate the transfer of genetic material.

Contact inhibition   The phenomenon whereby most normal cells stop dividing when surrounded by other similar cells. Contact inhibition is seen in scar formation and in the development of cell monolayers in tissue culture.

Continuous cell lines   Tissue cultures formed from transformed cells. Continuous cell lines are also called immortal cell lines.

Continuous cultures   Cultures in which microbes can be maintained in a steady physiological state. The most common types of continuous culture systems are the chemostat and the turbidostat.

Coremium   A group of hyphae, or sporophores, cemented together, that is generally raised into an upright position, and that supports spores. Also called a **synnema**.

Cowpox   A virus disease of cattle that is characterised by the appearance of pustular skin lesions.

Cristae   Involutions of the mitochondrial inner membrane that greatly increase its surface area.

Cystic fibrosis   A genetic disease in which the sufferer produces abnormal amounts of mucus, resulting in respiratory distress and digestive problems. Patients with cystic fibrosis are particularly prone to chest infections.

Cytochrome   A protein, containing haem, that acts as an electron carrier. Cytochromes constitute major components of the electron transport chain used in aerobic respiration.

Cytopathic effect   A morphological change within a cell that results from virus infection.

Cytoplasm   The protoplasm that lies within the cell membrane.

D value   The time required to reduce the viability of a population of microbes by 90% when held at a specified temperature and in a specified matrix. This is also called the **decimal reduction time**, since the number of viable cells is reduced to 10% of the original viable count.

Death phase   The phase of microbial growth in a batch culture during which the number of viable cells declines, usually in an exponential fashion.

Defective virus particle   A virus particle that is incapable of causing infection on its own.

Denaturation   Alteration of nucleic acid or protein that results in the loss of normal biological activity. Denatured proteins may retain their antigenic characteristics. A toxoid is a denatured toxin molecule that can be used to stimulate antibody production and hence immunity from the effects of the native toxin. Molecules are frequently denatured by the effects of heat or chemicals.

Dendrogram   A pictorial representation of the degree of relatedness of organisms, derived from a similarity matrix. Dendrograms have a tree-like appearance.

Dermatophyte   A fungus that parasitises skin and causes ringworm.

Deuteromycetes   Fungi Imperfecti. These are fungi in which a sexual reproductive phase has not been identified.

Dichroic   Showing two colours. A dichroic mirror will only reflect light of particular wavelengths.

Dikaryotic   Cells that contain two haploid nuclei that undergo simultaneous division during the formation of new cells.

Dimorphic fungus   A fungus that can exist either as a yeast or a mycelial form.

Diphtheria   Acute infection of mucous membranes particularly of the throat, and associated with the formation of a false membrane. Diphtheria is caused by toxigenic strains of *Corynebacterium diphtheriae*.

Dipicolinic acid   A compound found in large quantities in bacterial spores.

Diplococcus   (Plural: diplococci.) Cocci that, most often, occur in pairs.

Diploid   (of a eukaryotic nucleus.) Having pairs of chromosomes.

Disinfectant   Chemical used for the decontamination of an environment that acts by killing the vegetative cells of microorganisms.

Disinfection   The removal or inhibition of microorganisms that are likely to cause disease from an object or environment.

DNA integration   The process by which foreign DNA becomes associated with the genome of its host cell. Integration of DNA is common in tumour-inducing viruses and lysogenic bacteriophages.

DNase   An enzyme that breaks down DNA.

Dolipore septum   Septum with a narrow central pore flanked by cap-like flanges found in Basidiomycetes.

Doubling time   The time required for a microbial population to double in number or in biomass.

Durham tube   Small, inverted tube used in a broth culture to detect the formation of gas from the fermentation of carbohydrates.

Dysentery   Infection characterised by diarrhoea in which stools contain blood and mucus. Amoebic dysentery is caused by *Entamoeba histolytica* and bacilliary dysentery is caused by bacteria of the genus *Shigella*.

Early messenger RNA   Messenger RNA that is produced shortly after virus infection. Early messenger RNA often encodes proteins that are necessary for virus replication within the infected host cell.

Eclipse period   The time in a virus growth cycle during which the infectious virus particles cannot be recovered from within the infected cells. The eclipse period usually occurs shortly after the virus has penetrated the cell.

Elek test   Immunological test used in the identification of toxigenic strains of *Corynebacterium diphtheriae*. Filter paper soaked in antiserum is placed across a streak inoculum of the strain to be tested, and the culture is incubated. A positive result is indicated by the formation of precipitation lines where the toxin and antitoxin have diffused and interacted.

Embryo   A developing animal before birth or hatching.

Embryonated egg   A fertilised egg, in which an embryo is developing.

Endocarditis   Infection of the lining of the heart, particularly associated with heart valves.

Endocytosis   The process whereby foreign particles enter cells. Local invaginations of the cell membrane enclose the particle and its surrounding fluid, and the particle then becomes engulfed within its own vesicle in the cell cytoplasm.

Endogenous   Originating from or growing within a structure.

Endoplasmic reticulum   Complex membrane structure found within the cytoplasm of eukaryotic cells.

Endospore   A thick-walled spore formed within a sporangium. Endospores are highly resistant structures, affording protection against heat, irradiation and chemicals. They are often referred to simply as **spores**.

Endosymbiont   An organism that lives within another organism in a mutually beneficial association.

Endothelial cells   Cells that line the internal surfaces of organs.

Endotoxin   Lipopolysaccharide associated with the outer membrane of Gram-negative bacteria that is toxic. It is responsible for the symptoms of septic shock.

Enrichment media   Broth-based media that contain inhibitors, and that are designed to select for the growth of particular, desired microbes. These are inhibited to a lesser extent than other microbes in the original inoculum.

Enteroarthric conidium   A conidium that forms by fragmentation of part of a fungal hypha.

Enteroblastic conidium   A conidium that forms by swelling or budding from within a fungal hypha.

Enterotoxin   A toxin that causes the symptoms of enteritis, namely vomiting or diarrhoea.

Envelope   A membrane structure that surrounds the capsid of certain viruses. Envelopes contain lipids from the host cell, and protein encoded by the virus. Protein within virus envelopes may be present as lipoproteins as well as glycoproteins. Envelopes may be derived from the nuclear membrane or the plasma membrane of the host cell, depending upon the type of virus.

Enveloped virus   A virus in which the capsid is surrounded by an envelope.

Episome   A plasmid that can exist independently or that can become integrated into the chromosome.

Epithelial cells   Cells lining the outer surface of the skin or internal organs.

Eubacteria   One of the two major groups of bacteria. Eubacteria have structures that are considered typical of most bacteria including cell walls, where present, that contain peptidoglycan.

Eukaryote   An organism that has a nucleus that is separated from the cytoplasm by a nuclear membrane, and a cytoplasm that contains numerous organelles and a complex arrangement of intracellular membranes.

Exogenous   Originating from or growing outside a structure. Exogenous spores are found externally on spore-bearing structures.

Exon   The coding region of a eukaryotic gene.

Exosporium   A thin structure, composed of protein, lipid and carbohydrate, that encloses the entire spore structure in some bacterial species.

Exotoxin   A toxic protein, produced by the normal metabolic processes of a microbe, and often released into its environment.

Explant culture   A special type of tissue culture in which cells grow out from a block of tissue aseptically removed from its normal position, and incubated in tissue culture medium.

Exponential growth   The phase of a batch culture during which the population of microbes increases in an exponential fashion. During this growth phase the time taken for the population to double remains constant. It is also referred to as the **logarithmic growth phase**.

Factor VIII   A blood-clotting factor, not made by people suffering from haemophilia.

Facultative anaerobe   An organism that grows in the presence of oxygen as well as in its absence, but facultative anaerobes grow better when oxygen is present.

Fermentation   The metabolism of organic compounds to release energy without the use of oxygen. Organic compounds are used both as electron donors and electron acceptors.

Fibrin   An insoluble fibrous protein responsible for blood clotting, and formed from fibrinogen by the action of the enzyme thrombin. Fibrin can also be formed by the action of bacterial coagulase.

Fibrinogen   A soluble serum protein that is converted into fibrin during the clotting of blood.

Fibroblast   An irregularly-shaped cell distributed throughout the connective tissue of vertebrates, and that is responsible for the production of collagen fibres.

Filterable virus   An old term used to describe infectious particles that could pass through filters that retain small bacteria. The word 'filterable' has now been dropped, and these particles are simply referred to as viruses.

Fimbria   (Plural: fimbriae.) A fine hair-like appendage of Gram-negative bacteria that aids adhesion.

Fission yeast   A yeast that divides by splitting into two progeny cells in a manner somewhat similar to bacterial transverse binary fission, rather than by budding.

Flagellin   The protein that forms flagella.

Flagellum   (Plural: flagella.) A thin whip-like appendage of microorganisms that is responsible for motility.

Fluorochrome   A chemical that absorbs light of a short wavelength and emits light of a longer wavelength. Fluorochromes are used to convert ultraviolet light into visible light.

Foetus   A mammalian embryo in which the main features of the fully developed animal can be recognised.

Foetal tissue   Tissue derived from a foetus.

Forespore   A precursor structure of an endospore formed within the bacterial sporangium.

Fruiting body   A specialised spore-bearing structure.

Fungi Imperfecti   Fungi in which sexual reproduction has not been observed, and that thus cannot be assigned a full classification. Also referred to as **Deuteromycotina** or **Deuteromycetes**.

Fungicidal agent   An agent that kills fungi.

Fungus   (Plural: fungi.) A eukaryotic organism that possesses a cell wall, and that requires a supply of organic matter from which it derives energy.

Gametangium   (Plural: gametangia.) A structure that contains gametes, e.g. sexual organs in fungi.

Gamete   A sex cell. Two gametes fuse to form a zygote, from which a new individual organism can develop.

Gas gangrene   An anaerobic infection in which body tissue is rapidly killed, and the dead tissue breaks down forming gas bubbles. The commonest cause of gas gangrene is *Clostridium perfringens*.

Gas vacuole   A gas filled structure found in some cyanobacteria and other aquatic bacteria that enables the cell to float near to the surface of water.

Gelatin   A protein obtained from animal carcasses, and used for the determination of specific proteinases. Gelatin was used as a gelling agent in early bacteriological media.

Generalised transduction   Transfer of any part of the bacterial genome from a donor bacterium to a recipient by the accidental incorporation of bacterial DNA in a bacteriophage.

Gene therapy   Treatment of disorders by the introduction of copies of a normal gene or genes into those cells that lack such a gene, or that carry a defective copy of the gene. Early attempts are being made at present at using gene therapy to treat cystic fibrosis.

Genome   All the genetic material in a cell.

Genotype   The genetic constituents of an organism.

Geometric progression   A series of numbers in which each term is obtained by multiplying the preceding term by a constant factor. The commonest geometrical progression in microbiology is based on multiplication by 2 giving a progression that starts 1, 2, 4, 8, 16, 32, 64, 128, 256, 512, 1024 and so on.

Geotropism   Growth in a particular direction in response to a gravitational force.

Germ cell   The cell formed upon germination of a spore.

Germ tube   Hyphal protuberance growing out from yeast cells of *Candida albicans* when this fungus is incubated in serum.

Germicidal agent   An agent that kills microbes. This term is usually applied to those microbes that cause disease.

Germinant   A substance that stimulates the germination of a spore.

Gills   Radial plates on the undersides of the fruiting bodies of basidiomycete fungi bearing the hymenial layer.

Glucan   A polymer of glucose.

Glycocalyx   A network of polysaccharide found on the outside of bacterial cells.

Glycoprotein   A protein to which carbohydrate is attached. Glycoproteins are not formed by prokaryotic cells.

Golgi apparatus   A membranous organelle of eukaryotic cells associated with the export of cellular products.

Gonorrhoea   A sexually transmissible disease caused by *Neisseria gonorrhoeae*.

gp 41   A glycoprotein of human immunodeficiency virus that induces the fusion of the virion with the plasma membrane.

gp 120   A glycoprotein of the human immunodeficiency virus that attaches to the CD4 receptor of certain lymphocytes. This glycoprotein is involved in the initiation of cell infection by the human immunodeficiency virus, and is its major immunogen.

Gram-negative   A cell that is unable to retain a crystal violet–iodine complex when exposed to organic solvents such as acetone or alcohols. Such cells require counterstaining before they can be visualised in bright-field microscopy.

Gram-positive   A cell that retains a crystal violet–iodine complex when exposed to organic solvents such as acetone or alcohol. Such cells appear blue-black when viewed in bright-field microscopy.

Gram-variable   A cell in which part appears Gram-positive, and the rest appears Gram-negative when stained using Gram's stain.

Haem   A pigment that chaelates compounds such as iron, and found in association with proteins such as cytochromes and haemoglobin. Haem is an absolute growth requirement for the culture of *Haemophilus influenzae.*

Haemagglutinin   A glycoprotein spike situated in the surface of the envelope of influenza viruses. It attaches the virus to target cells to facilitate entry of the virus into its new host. In the laboratory, haemagglutinin can be recognised by its ability to attach to certain types of red blood cells. This phenomenon is useful in the laboratory identification and assay of influenza virus.

Haemolysin   A substance that causes the haemolysis and breakdown of red blood cells. Haemolysis may be complete, as with β-haemolysis, or partial, as with α-haemolysis.

Haemophilia   A disease in which patients cannot make certain blood clotting factors, and are thus prone to severe bleeding when exposed to the slightest injuries.

Haemorrhage   The escape of blood from blood vessels. This term especially refers to profuse loss of blood from blood vessels.

Halophile   An organism that can grow in high salt concentrations.

Haploid   (of a eukaryotic nucleus.) Having a single set of chromosomes rather than chromosomes in pairs such as are present in diploid cells.

HeLa cells   Transformed cells derived from cervical carcinoma tissue removed from Helen Lane.

Heterocyst   The anaerobic cell within a filament of vegetative cells of cyanobacteria in which nitrogen fixation occurs. Heterocysts lack photosynthetic pigment found in other cyanobacterial cells.

Heteroduplex   A double-stranded DNA structure in which each strand is from a different source.

Heteroploidy   The condition in which cells contain an abnormal number of chromosomes.

Heterothallic   A fungus in which two compatible thalli are required for sexual reproduction.

Heterotroph   An organism that requires one or more organic compounds to grow.

Hfr strain   A bacterial strain in which the fertility plasmid, F, is integrated into the chromosome, and that, in consequence, can readily transfer its genes into a suitable recipient bacterium in conjugation experiments.

Histone   A small basic protein that is found associated with the DNA of eukaryotic cells.

Holoarthric conidium   A conidium that arises as a result of the fragmentation of a complete segment of a fungal hypha.

Holoblastic conidium   A conidium that buds up away from its supporting hypha.

Holothallic conidium   A conidium that forms with disarticulation of a part of the hypha (usually the tip).

Homothallic fungus   A self-fertile fungus in which a single thallus produces both male and female gametes and can reproduce sexually.

HTST pasteurisation   High-temperature short-time pasteurisation. The material to be pasteurised is held at 71.7 °C for 15 seconds.

Hugh and Liefson test   A test used to determine whether glucose is oxidised or fermented.

Hydrocephalus   An excess of cerebrospinal fluid surrounding the brain.

Hymenium   (Plural: hymenia.) A well-defined layer, found in a basidiocarp in which the basidia of basidiomycete fungi are located.

Hypersensitivity   An allergic condition.

Hypha   (Plural: hyphae.) A filament that constitutes part of the fungal mycelium.

Icosahedral   A symmetrical form, adopted by certain virus capsids, characterised by 20 triangular faces, 30 edges and 12 corners.

Immortal cell lines   Tissue cultures of transformed cells that are capable of indefinite passage.

Immunisation   The act of creating immunity. Protection of the body from infection or the effects of microbial toxins.

Immunogen   A protein or carbohydrate, etc. that stimulates a cellular and humoral response in a foreign host. Humoral responses involve production of a specific antibody.

Impetigo   Superficial infection of the skin associated with copious pus formation, and frequently caused by either *Staphylococcus aureus* or *Streptococcus pyogenes*.

Inclusion body   Granule of organic or inorganic material, usually used as an energy store, and found in the bacterial cytoplasm.

Indole   Breakdown product from the metabolism of tryptophan by an enzyme complex known as 'tryptophanase'.

Inflammation   Changes in tissues that occur in response to injury or infection. The cardinal signs of inflammation include heat, swelling, pain and redness.

Influenza   An acute infection caused by the influenza virus. It is characterised by fever, muscle pains, headache, nausea and inflammation of the respiratory tract.

Inoculum   The material used to inoculate a culture.

Insertion sequence   The simplest type of mobile genetic element, capable of more or less random insertion into bacterial DNA.

Insertional mutation   A mutation caused by the insertion of a sequence of DNA into a gene. Typically these are caused by transposons or insertion sequences.

Inspissation   The process of holding media at 85 °C on each of three successive days to effect sterilisation.

Integration   See *DNA integration*.

Interrupted mating experiment   A mating experiment using an Hfr strain as the donor, in which matings are permitted to occur for various periods before selecting progeny. The order of appearance of genes in the recipient is a function of the time that the experiment has been allowed to continue, and this is directly proportional to the position of these genes on the donor chromosome. It takes about 100 minutes to transfer the whole of the *Escherichia coli* chromosome to a suitable recipient.

Intron   The non-coding region of a eukaryotic gene, spliced from the final messenger RNA.

in vitro   Literally meaning in glass. This term usually refers to experimental material in isolation from or outside of the whole animal or plant.

in vivo   Literally meaning in life. This term is usually applied to experiments or situations that involve the whole animal or plant.

Iodophore   A complex of iodine and a polymer that releases its iodine over a relatively long period.

Isotonic solutions   Solutions of equal osmotic strength.

Karyogamy   The fusion of nuclei during the sexual reproduction of fungi.

Keratin   Tough, fibrous protein found in skin, hair, nails, feathers, etc.

L-form   A naturally occurring bacterium that lacks peptidoglycan.

Lag phase   The time at the start of a batch culture when no growth occurs, but during which microbes adjust to their new environment. The lag phase ends when growth begins.

Lamella   (Plural: lamellae.) A structure formed by layers of membranes.

Lanceolate diplococcus   A diplococcus in which the pair of cells appear compressed, so as to look like lancets.

Late lactose fermenter   An organism that can only ferment lactose upon prolonged incubation. This property is useful in the identification of *Shigella sonnei*.

Late messenger RNA   Messenger RNA that is formed a considerable time after the initial virus infection of a cell. Late messenger RNA may encode the structural proteins of a virus.

Latent period   The time taken from the point of infection of a bacterial culture with a bacteriophage to the bursting of the first bacterium, resulting in the appearance of new virus particles in the growth medium.

Legionnaire's disease   Atypical pneumonia caused by *Legionella pneumophila* and named after the convention of American legionnaires at which the disease first became apparent.

Leprosy   Chronic infection caused by *Mycobacterium leprae* and affecting skin and nerves. Leprosy causes gross deformities of the tissues that are infected.

Leukocyte   A white blood cell.

Lipase   An enzyme that breaks down lipid to produce free fatty acids.

Lipopolysaccharide   A complex structure of sugar residues and fatty acids, particularly associated with the outer membrane of Gram-negative bacteria.

Lipoprotein   A complex of protein with lipid. Lipoprotein structures are important structural components of the cell walls of Gram-negative bacteria.

Lithotroph   An organism that obtains its energy from inorganic compounds.

Logarithmic growth   The phase of a batch culture during which the population of microbes increases in a logarithmic fashion. During this growth phase the time taken for the population to double remains constant. It is also referred to as the **exponential growth phase**.

Lophotrichous   Having a polar tuft of flagella.

LTH pasteurisation   Low-temperature holding pasteurisation. The material to be pasteurised is held at 62.8 °C for at least 30 minutes.

Luciferase   An enzyme that releases light as a result of the breakdown of ATP.

Lyme disease   Septic arthritis caused by *Borrelia burgdorferi*.

Lymphocyte   A white blood cell that is formed within lymphoid tissue.

Lysis   The rupture of a cell.

Lysogen   A bacterium that carries a prophage in a stable manner and can lyse to produce infectious bacteriophage particles under suitable conditions.

Lysogenic bacteriophage   A bacteriophage that can stably co-exist with its host bacterium as a prophage.

Lysosome   A membrane-bound eukaryotic organelle that contains hydrolytic enzymes.

Lysozyme   An enzyme that weakens bacterial cell walls. **Phage lysozyme** is a type of lysozyme that is produced by bacteriophages, and that facilitates bacteriophage infection of bacteria.

Lytic cycle   A cycle of bacteriophage growth that results in the bursting of the infected bacterium to release the progeny virus particles.

Macroconidium   A large asexual conidiospore formed by filamentous fungi.

Malaria   Intermittent fever caused by protozoan parasites of the genus *Plasmodium*.

Mannan   A polymer of the sugar mannose, found in the cell walls of yeasts.

Mastitis   Inflammation or infection of the mammary glands. This is a particular problem in dairy herds in which cows suffer infection of the udders.

Meiosis   The process of cell division in which diploid cells that carry pairs of chromosomes give rise to haploid cells that contain only a single chromosome of each type.

Meningitis   Inflammation or infection of the membranes that surround the brain and spinal cord. This may be caused by a wide variety of microbes, and may be a life-threatening infection.

Mesophile   A microorganism that typically grows in the temperature range 15–45 °C, and with a growth optimum of between 20–45 °C.

Mesosome   An invagination of the cell membrane of bacteria. Some people consider these structures to be artefacts of the preparation procedure used for microscopy.

Metabolism   All of the chemical reactions that occur within living cells. Almost all metabolic reactions are catalysed by enzymes.

Metachromatic granule   A granule of polyphosphate found in the cytoplasm of certain bacteria such as corynebacteria, and that stain a different colour when stained with basic blue dyes such as methylene blue. Metachromatic granules act as energy stores.

Methyl red test   A test used to determine whether bacteria produce organic acids as a result of the fermentation of glucose. If this is the case, then the pH of the medium falls below pH 4.4, and this is sufficient to turn the methyl red indicator red.

Microaerophile   An organism that is damaged by atmospheric concentrations of oxygen, but that requires a small amount of oxygen, typically between 2 and 5%, for growth.

Microconidium   A small asexual conidiospore formed by filamentous fungi.

Mitochondrion   (Plural: mitochondria.) A subcellular organelle of eukaryotic cells with a double membrane structure. Mitochondria are the site of aerobic respiration. The mitochondrial inner membrane houses the electron transport chain and ATP-generating system.

Mitosis   The process of eukaryotic cell division in which the two new nuclei contain the same number of chromosomes as was present in the nucleus of the parent cell.

Mixotroph   An organism that can either be autotrophic or heterotrophic depending upon the prevailing environmental conditions.

Monera   The kingdom that contains prokaryotic organisms.

Monocistronic messenger RNA   Messenger RNA of eukaryotic cells that contains only one recognisable site for ribosome attachment.

Monoclonal antibody   An antibody of a single type that is produced by a cell resulting from the fusion of an antibody producing cell and a type of cancer cell.

Monotrichous   Having a single flagellum.

Mordant   A substance used to fix dyes to cells and tissues.

Mould   A filamentous fungus.

Mucociliary escalator   The mechanism by which particles are removed from the lungs. Particulate matter that becomes entrapped in the mucus that lines the lungs is swept up, out of the lungs by the beating of cilia in the cells of the lining of the lung.

Mucopeptide   An alternative name for peptidoglycan, the structural polymer of the cell walls of eubacteria.

Murein   An alternative name for peptidoglycan.

Mushroom   The fruiting body of a basidiomycete fungus.

Mycelium   (Plural: mycelia.) The network of hyphae that constitutes the vegetative structure of moulds and of streptomycete bacteria.

Mycolic acid   A high molecular weight fatty acid with waxy properties. Mycolic acids are a major component of the cell walls of acid-fast and coryneform bacteria.

Mycorrhiza   A symbiotic association between a filamentous fungus and the roots of a plant. The plant provides the fungus with organic nutrients, and the fungus helps to provide a source of minerals for the plant.

***N*-acetylglucosamine**   One of the sugars found in peptidoglycan, and the monomeric unit of chitin.

N-acetylmuramic acid The sugar in peptidoglycan that has the oligopeptide attachment.

N-acetyltalosaminuronic acid The sugar that carries the oligopeptide cross-links in the cell walls of some archaebacteria.

Nagler test A test used to identify the α-toxin of *Clostridium perfringens*. Half of an egg yolk plate is spread with antitoxin, and a streak culture of the bacterium to be tested is applied across each half of the plate. On the side of the plate that does not have the antitoxin, the α-toxin will break down the lecithin in the medium forming a precipitate, whereas in the side of the plate that carries the antitoxin, no precipitate will form.

Nascent Newly formed.

Nephelometer An instrument used to enumerate microbes by detecting the scattering of light.

Neutrophile An organism that grows best at neutral pH.

Nicotinamide adenine dinucleotide (NAD) A co-enzyme that acts as an electron carrier in biosynthetic metabolism. NAD is an essential growth factor for bacteria in the genus *Haemophilus*.

Nitrogen fixation The formation of ammonia from atmospheric nitrogen. This capacity is a function of cyanobacteria, members of the genus *Rhizobium*, and a few other bacteria.

Nucleocapsid The central core of a virus particle composed of protein in close and intimate association with the virus nucleic acid.

Nucleoid The area in a bacterial cell that contains the chromosomal DNA.

Nucleotide sequence The order in which nucleotides appear in nucleic acids. It is the nucleotide sequence that carries the genetic code of the nucleic acid.

Nucleus The membrane-bound area of a eukaryotic cell that carries the chromosomes.

Obligate anaerobe An organism that is inhibited or killed by the presence of oxygen.

Obligate intracellular parasite A microorganism or virus that can only replicate inside living cells. Obligate intracellular parasites frequently harness much of the host cell machinery for their own replication.

Oligonucleotide probe A short sequence of nucleotides, typically between 10 and 20 bases in length, that is labelled in some way and is used to hybridise with its complementary nucleic acid sequence. Oligonucleotide probes are used to locate specific genes.

Oncogene A gene that is involved in the development of a cancer. Oncogenes may be of virus or cellular origin.

One-step growth curve An experiment used to study the replication of lytic bacteriophages, in which the virus causes lysis of all infected bacteria within a culture to release new bacteriophage particles.

Oogonium   (Plural: oogonia.) In lower fungi the female sex organ that contains one or more eggs.

Oospore   The spore that forms after the fertilisation of an egg within an oogonium.

Opsonin   A substance that makes material more amenable to phagocytosis. Complement and various antibodies are potent opsonins.

Optimum growth temperature   The temperature at which the growth rate of a microbe is greatest.

Organ tropism   The predisposition of certain viruses for infection of particular organs.

Organelle   A defined location within a cell that carries out a particular function.

Osmoprotectant   A substance that helps cells to withstand osmotic pressures.

Osmosis   The movement of water across a semi-permeable membrane from a dilute solution to a more concentrated solution.

Osmotic pressure   The force that builds up as water moves across a semi-permeable membrane as a result of osmosis.

Osteomyelitis   Infection of the bone. The long bones of the leg are most commonly affected in acute osteomyelitis, and the spine may be involved in chronic, tuberculous osteomyelitis.

Ostiole   The opening of the perithecium of an ascomycete fungus through which asci are released. Also found in pycnidia.

Outer membrane   A specialised membrane located beyond the peptidoglycan layer of the cell walls of Gram-negative bacteria. The outer membrane contains lipopolysaccharide, and this acts as an endotoxin.

Outer membrane protein   A protein that is located in the outer membrane of Gram-negative bacteria. Outer membrane proteins often function in the transport of materials into and out of the periplasmic space.

Overlay medium   A medium containing a low concentration of agar, typically 0.7%, or a viscous solution of cellulose derivatives such as methyl cellulose. Overlay media are poured over infected tissue cultures or bacterial lawns to prevent secondary plaque formation in plaque assays.

Oxidase test   A test used to detect the presence of cytochrome oxidase. A 1% solution of tetramethyl-*para*-phenylene diamine dihydrochloride rapidly turns an intense purple colour when in contact with cytochrome oxidase.

Palisade   An arrangement of bacilli, where cells are aligned like fence posts. This is a feature of corynebacteria.

Papain   An enzyme that breaks down proteins.

Parasite   An organism that derives its nutrients from a living plant or animal, often but not always to the detriment of its host.

Particle infectivity ratio   The ratio of the total number of virus particles to the number of infectious virus particles present in a sample.

Pasteurisation   The process of heating a substance to a controlled temperature for a specified time with the aim of killing the vegetative cells of potentially harmful or damaging microorganisms. This process is widely used in the food industry where exposure of the food or beverage to high temperatures will spoil the quality of the product.

Pathogen   An organism capable of causing disease.

Pathogenicity   The ability of a microbe to cause disease.

Pellicle   The film on the surface of a liquid growth medium formed as the result of the growth of microbes at the fluid–air interface.

Pepsin   An enzyme that breaks down protein.

Peptide   A chain of amino acid residues.

Peptidoglycan   The structural polymer found in bacterial cell walls. Peptidoglycan has a sugar backbone of alternating units of *N*-acetyl-glucosamine and *N*-acetylmuramic acid. The *N*-acetylmuramic acid residues are cross-linked with short chains of amino acid residues. The peptide chains in peptidoglycan are unique in biological material because they contain both L- and D-isomers of amino acids.

Peptone   Partially hydrolysed protein, much used in microbiological growth media.

Periplasmic space   The space between the outer membrane and the cytoplasmic membrane of Gram-negative bacteria.

Perithecium   (Plural: perithecia.) A type of ascocarp or fruiting body made by certain ascomycete fungi. A perithecium has an ostiole or opening through which ascospores are expelled.

Peritrichous   Having flagella that cover the entire surface of the cell.

Persistent virus infection   A virus infection that is not terminated by the normal host responses, and that does not result in the death of the host, at least in the medium term.

Petite   Yeast mutants which lack mitochondria and consequently can obtain energy only by fermentation. As a result the colonies they give rise to are considerably smaller than wild-type yeasts.

pH   A measure of the hydrogen ion concentration, and hence the acidity or alkalinity of a sample.

Phagocytosis   The process whereby a cell engulfs a foreign particle in a membrane-bound enclosure or phagosome.

Phagosome　A vesicle formed within a cell as a consequence of phagocytosis.

Phenetic classification　A system of classification based upon the mutual similarities of organisms.

Phenotype　The characteristics of an organism that result from expression of its genotype.

Photoautotroph　An organism that uses light as a source of energy and exploits carbon dioxide as its sole carbon source.

Photolithotroph　An organism that uses light energy and inorganic compounds for growth, and that uses carbon dioxide as its sole carbon source.

Photoorganotroph　An organism that uses light energy and requires a supply of organic compounds for growth.

Photosynthesis　The process whereby light energy is harvested and used to convert carbon dioxide and water into organic compounds.

Phototroph　An organism that uses light as a source of energy.

Phycomycetes　The 'lower' fungi.

Phyletic classification　A system of classification that is based upon the ancestral or evolutionary relationship of organisms rather than their mutual similarities. Phyletic classification is also referred to as **phylogenetic classification**.

Pileus　The cap of a basidiomycete fungus under which the gills develop.

Pilus　(Plural: pili.) Hair-like appendage of a bacterial cell. Sex pili join the donor and recipient bacterial cells together during conjugal mating.

Pinocytosis　The process whereby a cell membrane entraps a small volume of its surrounding fluid along with the substances that fluid holds to form an intracellular vacuole.

Plague　Generally fatal infection caused by *Yersinia pestis*. This infection is spread by fleas that feed on infected rodents.

Plaques　Clear zone in a lawn of bacteria where the cells have been destroyed by the activity of a bacteriophage or focus of infection within a tissue culture monolayer caused by the activity of a virus. In a cell monolayer, formation of a plaque may or may not result in a zone of clearing.

Plaque assay　An assay based on the ability of an infectious virus particle to produce a discrete focus of infection. Foci of infection produced in a plaque assay may be enumerated, and in this way the number of infectious virus particles may be determined. The number of such particles in a sample is called its titre.

Plasma　(of blood.) The fluid portion of unclotted blood.

Plasma　(of protoplasm.) Viscous substance forming the bulk of material within the cell.

Plasmalemma   The cell membrane, enclosing the cytoplasm and nuclear material.

Plasma membrane   The semi-permeable membrane that surrounds the cytoplasm of a cell.

Plasmid   DNA that can exist independently of the chromosome. Plasmids are generally stably inherited, and may be present as single copies or multiple copies within a cell. Plasmids may encode features that are not generally required for the survival of an organism, but that may give the host a selective advantage under special circumstances. For example, many genes that confer antibiotic resistance are plasmid encoded.

Plasmid incompatibility   The inability of plasmids of the same incompatibility type to co-exist in the same cell as one another.

Plasmogamy   The fusion of two protoplasts during the sexual reproduction of fungi.

Plasmolysis   Shrinkage of the contents of the cell as a result of the loss of water due to osmosis.

Plastid   Small bodies found in plant cells. Plastids are frequently pigmented. Chloroplasts are plastids that contain chlorophyll.

Pleomorphism   Displaying a variety of shapes.

Pneumonia   Infection of the substance of the lung. Lobar pneumonia involves the infection of one or more lobes of a lung. Bronchopneumonia involves a dispersed infection of the bronchial tree.

Pock assay   An assay performed using embryonated eggs from chickens. Some viruses can be grown in chicken egg membranes to produce pocks. These can be enumerated to determine the titre of such viruses.

Poliomyelitis   Infection of the nervous system that may result in the paralysis and wasting of certain muscle groups.

Polyfructan   A polymer of fructose, associated with the formation of dental plaque.

Polyglucan   Polymer of glucose associated with the formation of dental plaque.

Polymerase chain reaction   The reaction that is used to amplify the sequences of DNA by repeated replication using a pair of oligonucleotide primers that bind to either strand of the target sequence at each extremity. Repetitive replication of the DNA is achieved by the use of thermostable DNA polymerase isolated from the thermophilic bacterium *Thermus aquaticus*.

Polypeptide   A long chain of amino acid residues.

Polysaccharide   A long chain of sugars.

Polysome   A chain of ribosomes, lined up along a messenger RNA molecule, each with a newly forming peptide chain attached.

Porin   A protein that helps to form channels through the outer membrane of Gram-negative bacteria.

Primary cell lines   Tissue cultures, often containing mixed cell types, that are obtained when cells are first released from the tissue that is being cultured.

Prion   An infectious particle thought to cause diseases such as scrapie in sheep. They have a protein structure, but they have not been found to contain nucleic acid.

Prodigiosin   The red pigment produced by *Serratia marcescens*.

Progeny   Offspring, or descendants.

Prokaryote   An organism that lacks a membrane-bound nucleus, and that has a relatively simple cell architecture.

Prophage   The virus DNA of a lysogenic bacteriophage that becomes integrated into the DNA of its host.

Protease   An enzyme that breaks down protein.

Proteolysis   The breakdown of proteins, frequently catalysed by enzymes known as proteases or proteinases.

Proteose   A breakdown product of protein.

Protista   The kingdom that contains mainly unicellular eukaryotic organisms.

Protoplast   A cell that lacks its cell wall.

Protozoon   (Plural: protozoa.) Unicellular eukaryotic animal.

Pseudohypha   (Plural: pseudohyphae.) A fragile branched chain of elongated cells, usually formed by the budding of a yeast.

Pseudomurein   Structural polymer found in the cell wall of archaebacteria.

Pseudomycelium   (Plural: pseudomycelia.) A fragile branched chain of elongated cells, usually formed by the budding of a yeast.

Psychophile   An organism that is capable of growth at low temperature, having an optimum growth temperature of 15 °C or lower, and a maximum growth temperature of about 20 °C.

Psychrotroph   An organism growing at low temperature. Sometimes referred to as a facultative psychrophile.

Pus   The thick yellowish fluid that is produced in response to infection. It contains serum, leukocytes and dead cells.

Pustule   A skin lesion characterised by a pus-filled vesicle.

Pycnidium   (Plural: pycnidia.) A fruiting body that is often bottle- or flask-shaped, in which asexual conidia are formed.

Pyrimidine dimer   An aberrant structure formed in DNA in response to ultraviolet irradiation that results in mutation.

Pyrrole ring   A ring structure found in organic molecules such as indole.

Q fever   An influenza-like illness caused by *Coxiella burnetii*.

Quellung reaction   Swelling of a capsule in response to the presence of anti-capsular antibodies.

Receptor   A molecule or molecules on the surface of a cell that is involved in the attachment of a virus anti-receptor as part of the interaction that is required for cell infection.

Receptor-mediated endocytosis   One of the models for the infection of cells by viruses.

Recombinant DNA   A term to describe DNA that is produced by the techniques of genetic engineering. Typically, a desired target sequence of DNA is generated, and inserted into a DNA vector using DNA ligase, an enzyme that joins DNA sequences together.

Reference particles   Particles, typically latex beads, added in a known concentration to a virus sample, and used to enumerate the total number of virus particles in that sample, using electron microscopic techniques.

Replication   The term used to describe reproduction in viruses.

Repressor protein   A protein encoded by a regulatory gene that prevents transcription of a particular gene.

Respiration   An energy-yielding process involving the oxidation of a substrate such as glucose.

Respiratory transport chain   A series of compounds located within a membrane structure that transfer electrons from an oxidisable substrate to a terminal electron acceptor. Cytochromes are a principal component of the respiratory transport chain.

Restriction endonuclease   An enzyme that recognises a particular short sequence of DNA and cuts the double-stranded structure at, or very close to, that point. The recognition sites for restriction endonucleases are typically four or six base-pairs long.

Restriction endonuclease polymorphism   The difference in patterns seen following separation of DNA fragments by electrophoresis obtained when related, but not identical, sequences are digested with restriction endonucleases. These are also referred to as restriction fragment length polymorphism, or RFLP for short.

Reverse transcriptase   An enzyme that catalyses the synthesis of cDNA from an RNA template. Reverse transcriptase is associated with retroviruses such as human immunodeficiency virus.

Rhizoid   A tapering, root-like structure by which some primitive fungi are attached to their substrate.

Rhizomorph   A structure in which fungal hyphae aggregate together and become differentiated into a rope-like structure.

Ribosome   A subcellular structure made up of ribosomal RNA and protein that effects the translation of messenger RNA into protein.

Ringworm   A skin infection caused by a dermatophyte fungus.

Rise period   The period during a one-step growth experiment when infected cells lyse to release newly-formed virus particles.

Sabin vaccine   A vaccine that protects against poliomyelitis, and that contains live, attenuated virus particles.

Salk vaccine   A vaccine that protects against poliomyelitis, and that contains killed polio virus particles.

Sanitisation   The reduction of the microbial load in an environment to a level considered acceptable for the protection of public health.

Saprophyte   An organism that lives on dead or decaying organic matter.

Sarcoma   A malignant tumour derived from connective tissue cells.

Satellitism   The growth of *Haemophilus influenzae* around a colony of *Staphylococcus aureus* on fresh blood agar. *Haemophilus influenzae* requires both X- and V-factors to grow. Fresh blood agar contains a rich supply of X-factor, and the staphylococcal colony provides the V-factor, and therefore close to the staphylococcal colony, the haemophilus colonies grow large, but the further away they are from the staphylococcal colony, then the smaller the haemophilus colonies become.

Scalded skin syndrome   A skin infection caused by strains of *Staphylococcus aureus* that cause the reddening and sloughing of skin, causing it to take on a scalded appearance.

Scarlet fever   An infection caused by *Streptococcus pyogenes* that is characterised by a high fever, sore throat and a marked skin rash, from which the disease gets its name.

Sclerotium   (Plural: sclerotia.) A hardened food laden structure that enables certain moulds to survive in a dormant state.

Scrapie   The colloquial term for a neurological disorder of sheep characterised by an intense itching and thirst. At post mortem, the brain tissue is vacuolated, and it takes on the appearance of a sponge. Diseases such as scrapie are termed spongiform encephalopathies.

Secondary metabolite   A product of metabolism that is made after growth is completed. Antibiotics and many pigments are secondary metabolites.

Selective media   Solid media that incorporate inhibitors, and that are designed to select for the growth of particular, desired microbes. These are inhibited to a lesser extent than other microbes present in the original inoculum.

Semi-continuous cell lines Tissue cultures that are generally obtained from aborted foetal tissues, and with a greater capacity for maintaining cell divisions than tissue cultures obtained from tissues taken from adults.

Septic shock Clinical condition characterised by low blood pressure and a very high risk of organ failure. This is a consequence of the presence of microorganisms in the blood. Septic shock is a complication of septicaemia.

Septicaemia The clinical condition that is caused by the presence of microorganisms in the blood. The typical signs include fever, rapid breathing and chills.

Septum (Plural: septa.) A partition or wall.

Serial dilution The repeated dilution of a sample by transferring a fixed volume from one dilution into a fixed volume of diluent to prepare the next dilution. Common serial dilutions include doubling dilutions in which one unit of sample is repeatedly diluted with one unit of diluent, and decimal dilutions where one volume of sample is mixed with nine volumes of diluent to make up a total diluted volume of ten units.

Serological test A laboratory test that exploits the interaction between antibodies and antigens.

Serotyping The process of differentiating strains of microorganisms known as serovars on the basis of their antigenic structure. This is commonly achieved by agglutinating microbial cells with specific antisera.

Serovar A strain of organism that has been separated from other types on the basis of serological differences.

Serum The yellow fluid component of blood from which the clot and blood cells have been removed.

Sexual reproduction Reproduction involving the fusion of gametes (sex cells).

Similarity matrix A matrix used in numerical taxonomy in which rows and columns represent the two organisms being studied, and the values show the degree of similarity between the two organisms.

Slime layer A gelatinous covering around the outside of a cell. Slime layers are generally considered to be less dense than capsules, but sometimes the terms are used interchangeably.

Slime moulds Microbes that lack fungal cell walls, and that have an amoeboid, unicellular life-style except for reproduction, when cells congregate and develop into structures that possess fungus-like properties.

Smallpox A virus infection caused by variola virus, and characterised by high fever and a pustular skin rash. In unvaccinated populations, smallpox carried a very high risk of mortality.

Southern blotting The process of separating DNA according to its size using electrophoresis, transferring it to a suitable membrane support, and using a labelled DNA probe to determine the location of a desired gene or other DNA sequence. This technique is often named after its originator, Ed Southern.

sp. or spp.   Abbreviations for undesignated species following the name of a genus where 'sp.' indicates one undefined species and 'spp.' indicates more than a single undetermined species.

Specialised transduction   The transfer of bacterial genes from a donor to a recipient effected by a temperate bacteriophage. Only a very limited sequence of DNA can be transferred by such a bacteriophage because each temperate bacteriophage has a unique site of integration into its host genome.

Specific growth rate   The characteristic rate of growth of a microbe in a batch culture. The specific growth rate (μ) of an organism is related to its doubling time (t_d) by the formula $\mu = 0.69/t_d$

Spectrophotometer   An instrument that measures the absorbance of light.

Spheroplast   A Gram-negative bacterial cell with the cell wall removed but retaining its outer membrane.

Spoilage organisms   Microorganisms that bring about spoilage or decay of products, especially food.

Spongiform encephalopathy   A neurological disorder in which brain tissue degenerates, and at post mortem appears spongy, with large vacuoles present. These are thought to be caused by prions.

Sporangiophore   A structure that supports an asexual sporangium, produced in lower fungi.

Sporangiospore   A spore contained in a sporangium.

Sporangium   (Plural: sporangia.) A sac-like structure in which asexual spores are produced.

Spore   A resistant, resting structure formed by certain microorganisms.

Spore coat   Layer of a spore that protects the cortex. The spore coat is made up of layers of protein and inorganic phosphates, together with small amounts of lipid and polysaccharide.

Spore core   The central portion of the core, similar in composition to a dehydrated vegetative cell

Spore cortex   The layer of peptidoglycan that encloses the spore core and core wall of a bacterial spore.

Sporophore   (of fungi.) A general term for a spore-bearing structure in mycelial fungi.

Sporulation   The process of spore formation, sometimes referred to as **sporogenesis**.

Stationary phase   The phase of a batch culture when exponential growth is complete, and the number of new cells formed is balanced by the loss of cells from the culture, leaving the overall cell count constant.

Steady state growth   Growth in which the physiological state of the microbes in the culture remains constant.

Sterigma   (Plural: sterigmata.) A narrow, pointing stalk-like structure, which supports spores, e.g. conidia, basidiospores.

Sterilisation   The removal of all living cells, including spores, and viruses or other infectious agents from an object or environment.

Sterol   An organic chemical with three six-membered rings and a five-membered ring condensed together. Natural sterols generally include a long side-chain and an alcohol group. Sterols help in the stabilisation of eukaryotic membranes.

Stipe   The stalk of the fruiting bodies of basidiomycete fungi.

Streptomycetes   A group of branched bacteria. This group is responsible for the production of many clinically useful antibiotics.

Subunit vaccine   A nucleic acid-free vaccine composed of those parts of a virion that are immunogenic and, following inoculation, will provide protection against the whole virus particle. An example of a subunit vaccine is the hepatitis B surface antigen vaccine.

Sulphur granule   A granule of elemental sulphur found as an inclusion body in certain bacteria.

Supernatant   The fluid that lies above a pellet after separation of a mixture by centrifugation.

Superoxide dismutase (SOD)   An enzyme that catalyses the conversion of superoxide radicals into oxygen and hydrogen peroxide. The hydrogen peroxide produced by this reaction can be dissipated by the enzyme catalase, that converts it into water, releasing oxygen as well. Superoxide dismutase and catalase help to protect many types of cell from toxic forms of oxygen.

Superoxide radical   A radical formed by the addition of a single electron to molecular oxygen. Superoxide radicals are highly reactive, and very toxic to living cells.

Symbiosis   The association of dissimilar organisms where each partner derives benefit from the association.

Synchronous growth   The growth of a population in which all the cells divide at the same time.

Syphilis   Sexually transmissible disease that can also be passed on congenitally from mother to offspring, caused by *Treponema pallidum*.

Taxonomy   The study of biological classification. Taxonomy involves assigning organisms into groups, or the process of classification, the assignation of names to organisms, or nomenclature, and the identification of organisms.

Teichoic acid   A polymer of ribitol phosphate or glycerol phosphate occurring in the cell walls of certain Gram-positive bacteria. Teichoic acids act as antigens.

Teichuronic acid   A polymer similar to teichoic acid, but containing sugar acids such as *N*-acetylmannosuronic acid instead of phosphoric acids. Teichuronic acids are produced when Gram-positive bacteria are grown in conditions where phosphates are scarce.

Teleomorph   Sexual phase of a fungus.

Teliospore   A thick-walled resting spore, especially in lower basidiomycete fungi.

Temperate bacteriophage   A bacteriophage that is capable of stable existence in its bacterial host. Temperate bacteriophages form prophages that become integrated into the host genome, where they exist stably in a cell known as a lysogen. Under suitable conditions, the prophage in a lysogen is activated to replicate and produce infectious bacteriophages that burst from the host cell in a lytic process.

Temporal control   A series of events that occur in a timed sequence: for example the appearance of early and late messenger RNAs during a virus growth cycle.

Thallic conidium   A conidium that develops directly from fragmentation of, or separation from, a hypha.

Thallospore   A vegetative spore that is produced when a hypha becomes disjointed and fragments.

Thallus   The body of a fungus. Yeasts have a unicellular thallus, but moulds have a thallus that includes mycelia and their associated fruiting bodies.

Thermal death point   The lowest temperature at which a microbe is killed within a specified time and in a specific mix.

Thermal death time   The time taken to kill all of the microbes in a population held at a specified temperature and in a specific matrix.

Thermoacidophile   An organism that grows at an elevated temperature and at a low pH.

Thermophile   An organism that grows at an elevated temperature.

Thylakoid   The elaborate internal membrane system found within a chloroplast. This term is also used to describe the pigment-bearing membranes of photosynthetic bacteria.

Titre   A relative measure of the activity of a biological suspension, for example, a virus titre is an indication of the infectivity of a suspension of virus particles.

Toadstool   The fruiting body of a basidiomycete fungus, usually poisonous.

Tonoplast   The membrane that delineates the cell vacuole in a plant cell.

Toxic shock syndrome   Illness characterised by the sudden onset of a skin rash, fever, vomiting and diarrhoea. In severe forms of the disease the patient suffers low blood pressure, heart and kidney failure. Toxic shock syndrome is caused by strains of *Staphylococcus aureus* that produce toxic shock syndrome toxin 1, also called TSST-1.

Toxigenic   Capable of producing a toxin.

Toxin   A poison. In particular toxins are substances produced by organisms that in low concentrations can damage other organisms and cause disease.

Transconjugant   The product of a bacterial mating. Transconjugants are formed when suitable recipients acquire genetic material from donors as a consequence of bacterial conjugation.

Transcription   The production of single-stranded RNA using a DNA template. Exceptionally, some RNA viruses use RNA templates to produce single-stranded RNA.

Transduction   The process of transferring genetic material from one bacterium to another, using a bacteriophage as a vector.

Transformation   The transfer of genetic material in bacteria where transformation-competent bacterial cells take up naked DNA from their environment. See also *cell transformation*.

Translation   The process by which the genetic code of messenger RNA is used to direct the synthesis of new proteins with the aid of ribosomes, transfer RNA and other cellular components. Translation is commonly referred to as protein synthesis, although strictly this process requires both transcription and translation.

Transposable element   A sequence of DNA that can move from one location to an unrelated location within a genome.

Transposition   The process of relocation of a transposable element. This is sometimes referred to as **illegitimate recombination** since the transposable element does not need any sequence homology with its target sequence. Transposition events can lead to the formation of insertional mutants.

Transposon   A transposable element that has an identifiable genetic marker. Transposons often encode antibiotic resistance, but can also code for toxin production or the ability to utilise particular chemicals.

Transverse binary fission   The process of asexual reproduction that involves a single cell dividing transversely into two new cells.

Trichome   A chain of cells in which the individual members are intimately joined to form a hair-like structure.

Trypsin   An enzyme that breaks down protein.

Trypsinisation   The treatment of tissue cultures with trypsin to release individual cells during passaging.

Tryptone   The product of digestion of casein using trypsin. Tryptone is rich in tryptophan, and is used in microbiological growth media.

Tuberculosis   A chronic infection, frequently of the lungs but also found in other organs such as the kidneys or bone, and characterised by the presence of nodules called tubercles. Tuberculosis is caused by *Mycobacterium tuberculosis*, the bacterium that also causes tuberculous meningitis.

Tumour   Proliferation of a tissue as a result of unrestrained cell growth. Tumours may be benign, as are warts, or they may be malignant, such as those that form cancers.

Tumour cells   Cells derived from tumours, and which do not have the normal restrains operating upon cell division.

Turbidostat   A continuous culture apparatus in which the cell density is kept constant.

Tyndallisation   The process of holding a heat-sensitive medium at 90–100 °C for a ten minute period on each of three successive days to sterilise it.

Typhoid   A generalised infection caused by *Salmonella typhi*. Typhoid is characterised by fever, headache, constipation and muscle pains. Later in the disease the patient may suffer diarrhoea and intestinal haemorrhage.

Typhus   A generalised infection caused by *Rickettsia prowazekii*. It is characterised by general prostration and delirium, and the patient has a purple skin rash.

Ultraviolet irradiation   Irradiation with light of shorter wavelength than that seen in the visible spectrum.

Urease   The enzyme that splits urea to form carbon dioxide and ammonia. The ammonia that is released from this reaction increases the pH of the local medium.

V factor   An essential growth factor for bacteria of the genus *Haemophilus*. This has been identified as nicotinamide adenine dinucleotide (NAD).

Vaccination   Inoculation with a vaccine to provide immunity from an infectious agent or its toxins.

Vaccine   A preparation of killed or attenuated microorganisms, or the products of a microorganism that is used to induce active immunity against microbial disease.

Vacuole   A fluid-filled space. Certain aquatic organisms have gas-filled vacuoles to aid buoyancy.

Vector   A carrier of a disease or infection. This term also applies to structures that are used to introduce recombinant DNA into new hosts, so that they may express foreign genes. Plasmids, bacteriophages, vaccinia virus and adenoviruses have all been used as DNA vectors.

Vegetative reproduction (of fungi.) Reproduction involving the fragmenttaion and subsequent separate development of individual portions of the fungal thallus.

Vesicle   Small membrane-bound hollow body.

Vibrio   A bacillus that is slightly curved or comma-shaped.

Virion   An alternative name for a virus particle.

Viroid   An infectious agent of plants that consists of naked RNA with no protein associated.

Viricidal agent   An agent that can 'kill' viruses.

Virulence   The capacity to cause disease. Some authors use the term virulence as the degree to which a microbe can cause disease.

Virus   An infectious particle comprising nucleic acid in a protein coat. Some viruses may also have a lipid-containing envelope. All viruses are obligate intracellular parasites.

Virus envelope See *envelope.*

Virus growth cycle   The time from the attachment of a virus particle to its target cell to the release of progeny virions. The growth cycle of viruses can vary in length from less than one hour to several days.

Virus non-structural proteins   The virus-specific proteins that are found in virus infected cells that are not present in the virion. Most virus nonstructural proteins have an enzymatic function.

Virus structural proteins   Those proteins that are present in a virion.

Vitamin   An organic compound required in small amounts for growth and reproduction of an organism.

Voges–Proskauer test   A test for the production of acetoin from glucose.

Volutin granule   Granule of polyphosphate found in the cytoplasm of certain bacteria, and used as an energy store. See *metachromatic granules.*

Wild-type   The form of a gene found most commonly in organisms taken from their natural environment.

X-factor   An essential growth factor required by *Haemophilus influenzae*, and certain other species in the genus *Haemophilus.*

X-ray diffraction   A technique in which the diffraction pattern that crystals generate when placed in the path of an X-ray beam is analysed, in order to determine the structure of the crystalline solid.

Yeast   A predominantly unicellular fungus.

Yellow fever   An acute haemorrhagic fever that is associated with hepatitis, and inflammation of the kidneys. Yellow fever is a tropical disease, and is spread through mosquito bites.

Zoosporangium   (Plural: zoosporangia.) A sporangium in which asexual zoospores are produced.

Zoospore   A motile, flagellate, asexual spore found in lower, aquatic or soil-dwelling fungi.

Zygomycotina   Advanced   phycomycete   fungi.   Also   known   as **Zygomycota**.

Zygospore   A resting sexual spore that results from the fusion of two similar gametes in certain lower fungi.

Zygote   The diploid cell formed by the fusion of two haploid gametes.

Index